U.S.NRC

United States Nuclear Regulatory Commission

Protecting People and the Environment

NUREG-1907
Vol. 1

I0489277

Safety Evaluation Report

Related to the License Renewal of Vermont Yankee Nuclear Power Station

Docket No. 50-271

Entergy Nuclear Operations, Inc.

Manuscript Completed: May 2008
Date Published: May 2008

Office of Nuclear Reactor Regulation

ABSTRACT

This safety evaluation report (SER) documents the technical review of the Vermont Yankee Nuclear Power Station (VYNPS) license renewal application (LRA) by the United States (US) Nuclear Regulatory Commission (NRC) staff (the staff). By letter dated January 25, 2006, Entergy Nuclear Operations, Inc. (ENO or the applicant) submitted the LRA in accordance with Title 10, Part 54, of the *Code of Federal Regulations*, "Requirements for Renewal of Operating Licenses for Nuclear Power Plants." ENO requests renewal of the VYNPS operating license (Facility Operating License Number DPR-28) for a period of 20 years beyond the current expiration at midnight March 21, 2012.

VYNPS is located approximately five miles south of Brattleboro, Vermont. The NRC issued the VYNPS construction permit on December 11, 1967, and the operating license on February 28, 1973. VYNPS is of a Mark 1 Boiling Water Reactor (BWR) design. General Electric supplied the nuclear steam supply system and Ebasco originally designed and constructed the plant. The VYNPS licensed power output is 1912 megawatt thermal with a gross electrical output of approximately 650 megawatt electric.

This SER presents the status of the staff's review of information submitted through February 21, 2008, the cutoff date for consideration in the SER. The staff identified six confirmatory items which were resolved before the staff made a final determination on the LRA. SER Section 1.6 summarizes these items and their resolution. Section 6.0 provides the staff's final conclusion on the review of the VYNPS LRA.

Paperwork Reduction Act Statement

This NUREG contains information collection requirements that are subject to the Paperwork Reduction Act of 1995 (44 U.S.C. 3501 et seq.). These information collections were approved by the Office of Management and Budget (OMB), approval numbers 3150-0155; 3150-0011.

Public Protection Notification

The NRC may not conduct or sponsor, and a person is not required to respond to, a request for information or an information collection requirement unless the requesting documents display a currently valid OMB control number.

TABLE OF CONTENTS

Appendices

Tables

ABBREVIATIONS

AAC   alternate AC
AC   alternating current
ACAR   aluminum conductor alloy reinforced
ACI   American Concrete Institute
ACRS   Advisory Committee on Reactor Safeguards
ACS   alternate cooling system
ACSR   aluminum core steel reinforced
ADAMS   Agencywide Document Access and Management System
ADS   automatic depressurization system
AE   air evacuation
AEC   Atomic Energy Commission
AERM   aging effect requiring management
AFW   auxiliary feedwater
AISC   American Institute of Steel Construction
AM   aging management
AMP   aging management program
AMR   aging management review
ANSI   American National Standards Institute
AOG   augmented off-gas
APCSB   Auxiliary and Power Conversion Systems Branch
ART   adjusted reference temperature
AS   auxiliary system
ASME   American Society of Mechanical Engineers
ASTM   American Society for Testing and Materials
ATWS   anticipated transient without scram
AWWA   American Water Works Association

BAF   bottom of the active fuel
BLD   building drainage system
BOP   balance of plant
B&PV   Boiler and Pressure Vessel
BTP   Branch Technical Position
BWR   boiling water reactor
BWRVIP   Boiling Water Reactor Vessel and Internals Project

CAD   containment atmosphere dilution
CAP   corrective action program
CASS   cast austenitic stainless steel
CBI   Chicago Bridge & Iron
CCW   closed cooling water
CCWS   closed cooling water system
CD   condensate demineralizer
CDF   core damage frequency
CEA   control element assembly
CF   chemistry factor

CFR	*Code of Federal Regulations*
CI	confirmatory item
CLB	current licensing basis
CMAA	Crane Manufactures Association of America
CO_2	carbon dioxide
CPPU	constant pressure power uprate
CRL	component record list
CRD	control rod drive
CRGT	control rod guide tube
CS	core spray
CSS	core spray system
CSCS	core standby cooling system
CST	condensate storage and transfer
CUF	cumulative usage factor
CUFD	reactor water cleanup unit filter demineralizer
C_vUSE	Charpy upper-shelf energy
CW	circulating water
CWP	circulating water priming
DBA	design basis accident
DBE	design basis event
DC	direct current
DG	diesel generator
DLO	diesel lube oil
DW	demineralized water
ECCS	emergency core cooling system
EDG	emergency diesel generator
EFPD	effective full power days
EFPY	effective full-power year
EIC	electrical and instrumentation and control
EMPAC	Enterprise Maintenance, Planning, and Control
ENO	Entergy Nuclear Operations, Inc.
Entergy VY	Entergy Nuclear Vermont Yankee, LLC
EOL	end of life
EPRI	Electric Power Research Institute
EPRI-MRP	Electric Power Research Institute Materials Reliability Program
EPU	Extended Power Uprate
EQ	Environmental qualification
ER	Applicant's Environmental Report - Operating License Renewal Stage
ES	extraction steam
ESF	engineered safety feature
FAP	fatigue action plan
FAC	flow-accelerated corrosion
FCV	flow control valve
FW	feedwater
F_{en}	environmental fatigue life correction factor

FERC	Federal Energy Regulatory Commission
FF	fluence factor
FIV	flow-induced vibration
FO	fuel oil
FPC	fuel pool cooling
FPFD	fuel pool filter-demineralizer
FR	*Federal Register*
FSAR	final safety analysis report
ft-lb	foot-pound
FW	feedwater

GALL	Generic Aging Lessons Learned Report
GDC	general design criteria or general design criterion
GE	General Electric
GEIS	Generic Environmental Impact Statement
GL	generic letter
GSC	gland seal condenser
GSI	generic safety issue

HB	heating boiler
HCU	hydraulic control unit
HD	heater drain
HELB	high-energy line break
HPCI	high pressure coolant injection
HPCIS	high pressure coolant injection system
HPSI	high pressure safety injection
HVAC	heating, ventilation, and air conditioning
HV	heater vent
HWC	hydrogen water chemistry
HX	heat exchanger

I&C	instrumentation and controls
IA	instrument air
IASCC	irradiation assisted stress corrosion cracking
ID	inside diameter
IEEE	Institute of Electrical and Electronics Engineers
IGA	intergranular attack
IGSCC	intergranular stress corrosion cracking
IN	information notice
INEL	Idaho National Engineering Laboratory
INPO	Institute of Nuclear Power Operations
IPA	integrated plant assessment
IPE	individual plant examination
IR	insulation resistance
ISA	Instrument Society of America
ISG	interim staff guidance
ISI	inservice inspection

ISP	integrated surveillance program
IST	inservice testing
JDD	John Deere diesel
ksi	1000 pounds per square inch
KV or kV	kilo-volt
KW	kilo-watt
LOCA	loss of coolant accident
LPCI	low pressure coolant injection
LRA	license renewal application
LRBD	license renewal boundary drawings
LRIS	License Renewal Information System
LRPG	license renewal project guideline
MEB	metal-enclosed bus
MeV	mega-electron volt
MG	motor generator
MGLO	motor generator lube oil
MIC	microbiologically influenced corrosion
MS	main stream
MSIV	main stream isolation valve
MUD	make-up demineralizer
MWe	megawatts-electric
MWt	megawatts-thermal
N_2	nitrogen
NaOH	sodium hydroxide
NB	nuclear boiler
NBVIS	nuclear boiler vessel instrumentation system
n/cm^2	neutrons per square centimeter
NDE	nondestructive examination
NEI	Nuclear Energy Institute
NESC	National Electric Safety Code
NFPA	National Fire Protection Association
NPS	nominal pipe size
NRC	US Nuclear Regulatory Commission
NSAC	Nuclear Science Advisory Committee
NSSS	nuclear steam supply system
NUMARC	Nuclear Management and Resources Council (now NEI)
NUREG	US Nuclear Regulatory Commission Regulatory Guide
NUREG/CR	US Nuclear Regulatory Commission Regulatory Guide contractor report
NWC	normal water chemistry
ODSCC	outside-diameter stress corrosion cracking
OE	operating experience
OI	open item

PASS	post-accident sampling system
PCAC	primary containment atmosphere control
pH	potential hydrogen
P&ID	piping and instrumentation diagram
ppm	parts per million
P-T	pressure-temperature
PTS	pressurized thermal shock
PUSAR	power uprate safety analysis report
PVC	polyvinyl chloride
PW	potable water
PWR	pressurized water reactor
PWSCC	primary water stress corrosion cracking
QA	quality assurance
Q&A	question and answer
RAI	request for additional information
RBCCW	reactor building closed cooling water
RCIC	reactor core isolation cooling
RCPB	reactor coolant pressure boundary
RCS	reactor coolant system
RDW	radwaste
RFO	refueling outage
RG	regulatory guide
RHRS	residual heat removal system
RHRSW	residual heat removal service water
RIP	retired in place
RPV	reactor pressure vessel
RR	reactor recirculation
RRP	reactor recirculation pump
RRS	reactor recirculation system
RT	radiographic testing
RTD	resistance temperature detector
RT_{NDT}	reference temperature nil ductility transition
RV	reactor vessel
RVI	reactor vessel internals
RVID	reactor vessel integrity database
RWCU	reactor water cleanup
SA	service air
SBFPC	standby fuel pool cooling
SBGT	standby gas treatment
SBO	station blackout
SC	structure and component
SCC	stress-corrosion cracking
SE	safety evaluation
SER	safety evaluation report
SFP	spent fuel pool

SIF	stress intensification factor
SLC	standby liquid control
SO	seal oil
SPL	sampling
SOC	statement of consideration
SRP	Standard Review Plan
SRP-LR	Standard Review Plan for Review of License Renewal Applications for Nuclear Power Plants
SRV	safety relief valve
SS	stainless steel
SSC	system, structure, and component
SSE	safe-shutdown earthquake
SW	service water
SWS	service water systems
TBCCW	turbine building closed cooling water
TG	turbine generator
TLAA	time-limited aging analysis
TLO	turbine lube oil
TS	technical specifications
TRM	technical requirements manual
UFSAR	updated final safety analysis report
USAR	updated safety analysis report
USAS	United States of America Standard
USE	upper-shelf energy
UT	ultrasonic testing
UV	ultra violet
VHS	Vernon Hydroelectric Station
VT	visual testing
VYNPS	Vermont Yankee Nuclear Power Station
1/4 T	one-fourth of the way through the vessel wall

SECTION 1

INTRODUCTION AND GENERAL DISCUSSION

1.1 Introduction

This document is a safety evaluation report (SER) on the license renewal application (LRA) for Vermont Yankee Nuclear Power Station (VYNPS), as filed by Entergy Nuclear Operations, Inc. (ENO or the applicant). By letter dated January 25, 2006, ENO submitted its application to the United States (US) Nuclear Regulatory Commission (NRC) for renewal of the VYNPS operating license for an additional 20 years. The NRC staff (the staff) prepared this report to summarize the results of its safety review of the LRA for compliance with Title 10, Part 54, of the *Code of Federal Regulations*, "Requirements for Renewal of Operating Licenses for Nuclear Power Plants" (10 CFR Part 54). The NRC project manager for the license renewal review is Jonathan Rowley. Mr. Rowley may be contacted by telephone at 301-415-4053 or by electronic mail at JGR@nrc.gov. Alternatively, written correspondence may be sent to the following address:

Division of License Renewal
US Nuclear Regulatory Commission
Washington, DC 20555-0001
Attention: Jonathan Rowley, Mail Stop 011-F1

In its January 25, 2006 submission letter, the applicant requested renewal of the operating license issued in accordance with Section 104b (Operating License No. DPR-28) of the Atomic Energy Act of 1954, as amended, for VYNPS for a period of 20 years beyond the current expiration at midnight March 21, 2012. VYNPS is located approximately five miles south of Brattleboro, Vermont. The NRC issued the VYNPS construction permit on December 11, 1967, and the operating license on February 28, 1973. VYNPS is of a Mark 1 Boiling Water Reactor (BWR) design. General Electric supplied the nuclear steam supply system (NSSS) and Ebasco originally designed and constructed the plant. The VYNPS licensed power output is 1912 megawatt thermal with a gross electrical output of approximately 650 megawatt electric. The updated final safety analysis report (UFSAR) contains details of the plant and the site.

The license renewal process consists of two concurrent reviews, a technical review of safety issues and an environmental review. The NRC regulations in 10 CFR Part 54 and 10 CFR Part 51, "Environmental Protection Regulations for Domestic Licensing and Related Regulatory Functions," respectively, set forth requirements for these reviews. The safety review for the VYNPS license renewal is based on the applicant's LRA and responses to staff requests for additional information. The applicant supplemented the LRA and provided clarifications through its responses to the staff's requests for additional information in audits, meetings, and docketed correspondence. Unless otherwise noted, the staff reviewed and considered information submitted through February 21, 2008. The staff reviewed information received after that date case by case depending on the stage of the safety review and the volume and complexity of the information. The public may view the LRA and all pertinent information and materials, including the UFSAR, at the NRC Public Document Room, on the first floor of One White Flint North, 11555 Rockville Pike, Rockville, MD 20852-2738

(301-415-4737 / 800-397-4209), and at Dickinson Memorial Library, 115 Main St., Northfield, MA 01360. In addition, the public may find the LRA, as well as materials related to the license renewal review, on the NRC web site at http://www.nrc.gov.

This SER summarizes the results of the staff's safety review of the LRA and describes the technical details considered in evaluating the safety aspects of the unit's proposed operation for an additional 20 years beyond the term of the current operating license. The staff reviewed the LRA in accordance with the NRC regulations and the guidance in the US NRC NUREG-1800, Revision 1, "Standard Review Plan for Review of License Renewal Applications for Nuclear Power Plants" (SRP-LR), dated September 2005.

SER Sections 2 through 4 address the staff's evaluation of license renewal issues considered during the review of the LRA. SER Section 5 is reserved for the report of the Advisory Committee on Reactor Safeguards (ACRS). SER Section 6 presents the conclusions of this report.

SER Appendix A is a table of the applicant's commitments for renewal of the operating license. SER Appendix B is a chronology of the principal correspondence between the staff and the applicant on the LRA review. SER Appendix C is a list of principal contributors to this SER. Appendix D is a bibliography of the references in support of the staff's review.

In accordance with 10 CFR Part 51, the staff prepared a plant-specific supplement to NUREG-1437, "Generic Environmental Impact Statement for License Renewal of Nuclear Plants (GEIS)." This supplement discusses the environmental considerations related to the VYNPS license renewal. The staff issued a plant-specific supplement to the GEIS, "Generic Environmental Impact Statement for License Renewal of Nuclear Plants Supplement 30 Regarding Vermont Yankee Nuclear Power Station," on August 1, 2007.

1.2 License Renewal Background

Pursuant to the Atomic Energy Act of 1954, as amended, and NRC regulations, operating licenses for commercial power reactors are issued for 40 years. These licenses can be renewed for up to 20 additional years. The original 40-year license term was selected on the basis of economic and antitrust considerations, rather than on technical limitations; however, some individual plant and equipment designs may have been engineered based on an expected 40-year service life.

In 1982, the staff anticipated interest in license renewal and held a workshop on nuclear power plant aging. This workshop led the NRC to establish a comprehensive program plan for nuclear plant aging research. From the results of that research, a technical review group concluded that many aging phenomena are readily manageable and pose no technical issues for life extension of nuclear power plants. In 1986, the staff published a request for comment on a policy statement that would address major policy, technical, and procedural issues related to license renewal for nuclear power plants.

In 1991, the staff published the license renewal rule in 10 CFR Part 54 (Volume 56, page 64943, of the *Federal Register* (56 FR 64943), dated December 13, 1991). The staff participated in an industry-sponsored demonstration program to apply 10 CFR Part 54 to a pilot plant and to gain experience necessary to develop implementation guidance. To establish a scope of review for license renewal, 10 CFR Part 54 defined age-related degradation unique to license renewal. However, during the demonstration program, the staff found that many aging effects on plant systems and components are managed during the period of initial license. In addition, the staff found that the scope of the review did not allow sufficient credit for existing programs, particularly the implementation of 10 CFR 50.65, which also manages plant-aging phenomena. As a result, the staff amended 10 CFR Part 54 in 1995. As published in 60 FR 22461, dated May 8, 1995, the amended 10 CFR Part 54 establishes a regulatory process that is simpler, more stable, and more predictable than the previous 10 CFR Part 54 process. In particular, as amended, 10 CFR Part 54 focuses on the management of adverse aging effects rather than on identifying age-related degradation unique to license renewal. The staff initiated these rule changes to ensure that important systems, structures, and components (SSCs) will continue to perform their intended functions during periods of extended operation. In addition, the revised 10 CFR Part 54 rule clarifies and simplifies the integrated plant assessment for consistency with the revised focus on passive, long-lived structures and components (SCs).

In parallel with these initiatives, the NRC pursued a separate rulemaking effort (61 FR 28467, dated June 5, 1996) and developed an amendment to 10 CFR Part 51 to focus the scope of the review of license renewal environmental impacts and to fulfill the NRC's responsibilities in accordance with the National Environmental Policy Act of 1969.

1.2.1 Safety Review

License renewal requirements for power reactors are based on two key principles:

(1) The regulatory process is adequate to ensure that the licensing bases of all currently operating plants maintain an acceptable level of safety, with the possible exception of the detrimental aging effects on the functions of certain SSCs, as well as a few other safety-related issues, during the period of extended operation.

(2) The plant-specific licensing basis must be maintained during the renewal term in the same manner and to the same extent as during the original licensing term.

In implementing these two principles, 10 CFR 54.4, "Scope," defines the scope of license renewal as including those SSCs that (1) are safety-related, (2) the failure of which could affect safety-related functions, or (3) are relied on for compliance with the NRC fire protection, environmental qualification (EQ), pressurized thermal shock (PTS), anticipated transient without scram (ATWS), and station blackout (SBO) regulations.

Pursuant to 10 CFR 54.21(a), a license renewal applicant must review all SSCs within the scope of 10 CFR Part 54 to identify SCs subject to an aging management review (AMR). SCs subject to an AMR perform an intended function without moving parts or without a change in configuration or properties and are not subject to replacement after a qualified life or specified time period. As required by 10 CFR 54.21(a), license renewal applicants must demonstrate that the aging effects will be managed so that the intended function(s) of those SCs will be maintained consistent with the current licensing basis (CLB) for the period of extended operation. However, active equipment is considered to be adequately monitored and maintained by existing programs. In other words, detrimental aging effects that may affect active equipment are readily detectable and can be identified and corrected through routine surveillance, performance monitoring, and maintenance. Surveillance and maintenance programs for active equipment, as well as other maintenance aspects of plant design and licensing basis, are required throughout the period of extended operation.

Pursuant to 10 CFR 54.21(d), the LRA is required to include a UFSAR supplement that must have a summary description of the applicant's programs and activities for managing aging effects and an evaluation of time-limited aging analyses (TLAAs) for the period of extended operation.

License renewal also requires TLAA identification and updating. During the plant design phase, certain assumptions were made about the length of time the plant can operate. These assumptions were incorporated into design calculations for several plant SSCs. In accordance with 10 CFR 54.21(c)(1), the applicant must either show that these calculations will remain valid for the period of extended operation, project the analyses to the end of the period of extended operation, or demonstrate that the aging effects on these SSCs will be adequately managed for the period of extended operation.

In 2001, the NRC developed and issued Regulatory Guide (RG) 1.188, "Standard Format and Content for Applications to Renew Nuclear Power Plant Operating Licenses." This RG endorses Nuclear Energy Institute (NEI) 95-10, Revision 3, "Industry Guideline for Implementing the Requirements of 10 CFR Part 54 - The License Renewal Rule," issued in March 2001. NEI 95-10 details an acceptable method of implementing 10 CFR Part 54. The staff also used the SRP-LR in reviewing the LRA.

In the LRA, the applicant fully utilized the process defined in NUREG-1801, Revision 1, "Generic Aging Lessons Learned (GALL) Report," dated September 2005. The GALL Report summarizes staff-approved aging management programs (AMPs) for the aging of many SCs subject to an AMR. If an applicant commits to implementing these staff-approved AMPs, the time, effort, and resources to review the LRA can be greatly reduced, improving the efficiency and effectiveness of the license renewal review process. The GALL Report summarizes the aging management evaluations, programs, and activities credited for managing aging for most SCs throughout the industry. The report is also a quick reference for both the applicant and staff reviewers to AMPs and activities that can provide adequate aging management during the period of extended operation.

1.2.2 Environmental Review

Part 51 of 10 CFR governs environmental protection regulations. In December 1996, the staff revised the environmental protection regulations to facilitate the environmental review for license renewal. The staff prepared the Generic Environmental Impact Statement (GEIS) to document its evaluation of the possible environmental impacts of nuclear power plant license renewals. For certain environmental impacts, the GEIS establishes findings applicable to all nuclear power plants. These generic findings are codified in Appendix B, "Environmental Effect of Renewing the Operating License of a Nuclear Power Plant," to Subpart A, "National

1-4

Environmental Policy Act - Regulations Implementing Section 102(2)," of 10 CFR Part 51. Pursuant to 10 CFR 51.53(c)(3)(i), license renewal applicants may incorporate these generic findings in their environmental reports. In accordance with 10 CFR 51.53(c)(3)(ii), an environmental report must also include analyses of environmental impacts that must be evaluated on a plant-specific basis (i.e., Category 2 issues).

In accordance with the National Environmental Policy Act of 1969 and 10 CFR Part 51, the staff reviewed the plant-specific environmental impacts of license renewal, including whether the GEIS had not considered new and significant information. As part of its scoping process, the staff held a public meeting on June 7, 2006, in Brattleboro, Vermont, to identify plant-specific environmental issues. Draft, plant-specific GEIS Supplement 30 documents the results of the environmental review and makes a preliminary recommendation as to the license renewal action. The staff held another public meeting on January 31, 2007, in Brattleboro, Vermont, to discuss draft, plant-specific GEIS Supplement 30.

1.3 Principal Review Matters

Part 54 of 10 CFR describes the requirements for renewing operating licenses for nuclear power plants. The staff's technical review of the LRA was in accordance with NRC guidance and the requirements of 10 CFR Part 54. Section 54.29, "Standards for Issuance of a Renewed License," of 10 CFR sets forth the standards for license renewal. This SER describes the results of the staff's safety review.

In accordance with 10 CFR 54.19(a), the NRC requires license renewal applicants to submit general information. The applicant provided this general information in LRA Section 1. The staff reviewed LRA Section 1 and finds that the applicant has submitted the information required by 10 CFR 54.19(a).

In accordance with 10 CFR 54.19(b), the NRC requires that LRAs include "conforming changes to the standard indemnity agreement, 10 CFR 140.92, Appendix B, to account for the expiration term of the proposed renewed license." On this issue, in the LRA, the applicant stated:

> The agreement shall terminate at the time of expiration of the license specified in Item 3 of the attachment to the agreement, which is the last to expire. Item 3 of the attachment to the indemnity agreement, as revised by Amendment No. 6, lists VYNPS operating license number DPR-28. ENO requests that conforming changes be made to Article VII of the indemnity agreement, and Item 3 of the attachment to that agreement, specifying the extension of agreement until the expiration date of the renewed VYNPS facility operating license sought in this application. In addition, should the license number be changed upon issuance of the renewal license, ENO requests that conforming changes be made to Item 3 of the attachment and other sections of the indemnity agreement as appropriate.

The staff intends to maintain the original license number upon issuance of the renewed license, if approved. Therefore, conforming changes to the indemnity agreement need not be made and the requirements of 10 CFR 54.19(b) have been met.

In accordance with 10 CFR 54.21,"Contents of Application - Technical Information," the NRC requires that LRAs contain (a) an integrated plant assessment, (b) a description of any current licensing basis (CLB) changes occurring during the staff's review of the LRA, (c) an evaluation of TLAAs, and (d) a UFSAR supplement. LRA Sections 3 and 4 and Appendix B address the license renewal requirements of 10 CFR 54.21(a), 10 CFR 54.21(b), and 10 CFR 54.21(c). LRA Appendix A satisfies the license renewal requirements of 10 CFR 54.21(d).

In accordance with 10 CFR 54.21(b), the NRC requires that each year following submission of the LRA and at least three months before the scheduled completion of the staff's review, the applicant submit an LRA amendment identifying any CLB changes of the facility that materially affect the contents of the LRA, including the UFSAR supplement.

In accordance with 10 CFR 54.22, "Contents of Application - Technical Specifications," the NRC requires that the LRA include changes or additions to the technical specifications necessary to manage the aging effects during the period of extended operation. In LRA Appendix D, the applicant stated that it had not identified any technical specification changes necessary to support issuance of the renewed VYNPS operating license. This statement adequately addresses the 10 CFR 54.22 requirement.

The staff evaluated the technical information required by 10 CFR 54.21 and 10 CFR 54.22 in accordance with NRC regulations and SRP-LR guidance. SER Sections 2, 3, and 4 document the staff's evaluation of the technical information in the LRA.

As required by 10 CFR 54.25, "Report of the Advisory Committee on Reactor Safeguards," the ACRS will issue a report documenting its evaluation of the staff's LRA review and SER. SER Section 5 will incorporate the ACRS report when issued. SER Section 6 will document the findings required by 10 CFR 54.29.

The final, plant-specific GEIS Supplement 30 will document the staff's evaluation of the environmental information required by 10 CFR 54.23, "Contents of Application - Environmental Information," and will specify the considerations related to the VYNPS operating license renewal. The staff will prepare this supplement separately from the SER.

1.4 Interim Staff Guidance

License renewal is a living program. The staff, industry, and other interested stakeholders gain experience and develop lessons learned with each renewed license. The lessons learned address the staff's performance goals of maintaining safety, improving effectiveness and efficiency, reducing regulatory burden, and increasing public confidence. Interim staff guidance (ISG) is documented for use by the staff, industry, and other interested stakeholders until incorporated into such license renewal guidance documents as the SRP-LR and the GALL Report.

Table 1.4-1 shows the current set of interim staff guidance (ISGs), as well as the SER sections in which the staff addresses them.

Table 1.4-1 Current Interim Staff Guidance

ISG Issue (Approved ISG Number)	Purpose	SER Section
Nickel-alloy components in the reactor coolant pressure boundary (LR-ISG-19B)	Cracking of nickel-alloy components in the reactor pressure boundary. ISG under development. NEI and EPRI-MRP will develop an augmented inspection program for GALL AMP XI.M11-B. This AMP will not be completed until the NRC approves an augmented inspection program for nickel-alloy base metal components and welds as proposed by EPRI-MRP.	Not applicable [Pressurized Water Reactors (PWRs)only]
Corrosion of drywell shell in Mark I containments (LR-ISG-2006-01)	To address concerns related to corrosion of drywell shell in Mark I containments.	3.5.2.2.1

1.5 Summary of Open Items

As a result of its review of the LRA, including additional information submitted to the staff through July 3, 2007, the staff determined that no open items exist which would require a formal response from the applicant. An item would have been considered open if the applicant had not presented a sufficient basis for resolution of an issue.

1.6 Summary of Confirmatory Items

As a result of its review of the LRA, including additional information submitted to the staff through March 23, 2007, the staff identified the following confirmatory items (CIs). An item was considered confirmatory if the staff and the applicant had reached a satisfactory resolution, but the resolution had not been submitted to the staff. Each CI was assigned a unique identifying number. By letters dated July 3, July 30, and August 16, 2007, the applicant responded to these CIs. The staff reviewed these responses and closed each of the CIs. The basis for closing the CIs is as follows:

CI 2.3.3.2a-1

License renewal drawing LRA-G-191159-SH-01-0, at location H-11, depicts pipe section 2"-SW-566C as within the scope of license renewal. The license renewal boundary flag for 2"-SW-566C is located on an unisolable section of pipe. The actual location of the license renewal scope boundary for this pipe section is not clear. The staff requested that the NRC Regional Inspection Team perform an inspection to ensure that the license renewal scope boundaries for these components meet the requirements of 10 CFR 54.4(a)(2).

In Inspection Report 05000271/2007006, Vermont Yankee Nuclear Power Station - NRC License Renewal Inspection Report, dated June 4, 2007, Attachment, Review of Safety Evaluation Report Confirmatory Items, the regional inspection team stated in part that the applicant has included in-scope for spatial interaction the portion of the SW system in the service water pump area of the intake structure and the reactor building. Pipe section 2" SW-566C is in the reactor building and is therefore in-scope for spatial interaction. As described in LRA Section 2.1.2.1.3, portions of systems included for 10 CFR 54.4(a)(2) are not shown on LRA drawings. Further, the applicant's letter to the NRC dated July 3, 2007, LRA Amendment 27, Attachment 2 indicates that pipe section 4" SW-567 which attaches to pipe section 2" SW-566C is in-scope for spatial interaction.

Based on its review, the staff found the above response acceptable because the inspection team and the applicant acknowledged that service water pipe 2" SW-566C is within the scope of license renewal and subject to an AMR based on the potential for physical interaction with safety-related systems in accordance with 10 CFR 54.4(a)(2). Therefore, the staff concern described in CI 2.3.3.2a-1 is resolved.

CI 2.3.3.2a-2

LRA Section 2.1.2.1.2 states in part that nonsafety-related piping systems connected to safety-related systems were included up to the structural boundary or to a point that includes an adequate portion of the nonsafety-related piping run to conservatively include the first seismic or equivalent anchor. In addition, if isometric drawings were not readily available to identify the structural boundary, connected lines were included to a point beyond the safety/nonsafety interface, like a base-mounted component, flexible connection, or the end of a piping run (i.e, a drain line).

It is not clear whether the nonsafety-related piping systems were included up to the structural boundary or to a point that includes an adequate portion of the nonsafety-related piping run to include the first seismic or equivalent anchor. The staff requested that the NRC Regional Inspection Team perform an inspection to ensure that the license renewal scope boundaries for these components satisfy the requirements of 10 CFR 54.4(a)(2).

In Inspection Report 05000271/2007006, Vermont Yankee Nuclear Power Station - NRC License Renewal Inspection Report, dated June 4, 2007, Attachment, Review of Safety Evaluation Report Confirmatory Items, the NRC Regional Inspection Team stated in part that for structural support considerations, the applicant has included components outside the safety class pressure boundary, yet relied upon to provide structural/seismic support for the pressure boundary. The application describes the types of components which are included in the scope of license renewal for 10 CFR 54.4(a)(2) and subject to an AMR in the service water system in LRA Table 2.3.3-13-42. This table was developed by including all nonsafety-related portions of fluid systems which are located within a building containing safety-related components and all nonsafety-related piping connected to safety-related systems back to the structural boundary using an isometric drawing. In cases where an isometric drawing which depicts the structural boundary is not readily available, connected lines were included back to a point beyond the safety/nonsafety interface to a base-mounted component, flexible connection, or the end of a piping run (such as a drain line) in accordance with the response to RAI 2.1-2. As described in LRA Section 2.1.2.1.3, portions of systems included for 10 CFR 54.4(a)(2) are not shown on LRA drawings.

Further, the applicant's letter to the NRC dated July 3, 2007, LRA Amendment 27, Attachment 2 states that there are no nonsafety-related systems for which the applicant has not identified the nonsafety-related portions of systems which are attached to safety-related systems and required to be in the scope of license renewal in accordance with 10 CFR 54.4(a)(2). However, as a result of discussions with the staff during the Region I inspection (February 2007), the applicant determined that some safety-related SSCs in the VY turbine building required consideration for potential spatial impacts from nonsafety-related SSCs based on 10 CFR 54.4(a)(2). Therefore, an expanded review for SSCs in the turbine building determined that additional components required an AMR. Those additional component types have been added to LRA Table 2.3.3-13-42, as addressed in the applicant's letters to the NRC dated July 30, 2007 and August 16, 2007.

Based on its review, the staff finds the response acceptable because the NRC Regional Inspection Team found there are no nonsafety-related portions of systems which are attached to safety-related systems that are not within the scope of license renewal in accordance with 10 CFR 54.4(a)(2). Furthermore, the staff again reviewed the applicable LRA drawings for component types that may have been omitted from Table 2.3.3-13-42 and found all component types in Table 2.3.3-13-42 to be consistent with the component types included within the scope of license renewal at similar facilities. Therefore, the staff concern described in CI 2.3.3.2a-2 is resolved.

CI 2.3.3.12-1

LRA Section 2.3.3.12 indicates that the John Deere Diesel (JDD) is installed in compliance with 10 CFR 50, Appendix R, requirements. However, due to a lack of available drawings and/or detailed description of the diesel equipment listed in LRA Table 2.3.3-12, it is difficult to determine if any AMR category components may have been omitted from the table. It is recommended that the JDD be inspected to assure all AMR category components are included in the list of LRA Table 2.3.3-12. The staff requested that the NRC Regional Inspection Team perform an inspection to ensure that the license renewal scope boundaries for these components satisfy the requirements of 10 CFR 54.4(a)(3).

In Inspection Report 05000271/2007006, Vermont Yankee Nuclear Power Station - NRC License Renewal Inspection Report, dated June 4, 2007, Attachment, Review of Safety Evaluation Report Confirmatory Items, the NRC Regional Inspection Team stated that the John Deere diesel system components are listed in LRA Table 2.3.3-12 and the supporting fuel oil day tank, fiberglass underground storage tank, and supply lines are listed in LRA Table 2.3.3-6, "Fuel Oil System."

Based on its review, the staff found the above response acceptable because the NRC Regional Inspection Team verified that all components subject to an AMR are included in LRA Table 2.3.3-12 and LRA Table 2.3.3-6 and confirmed that no other portions of the John Deere diesel system should have been included within scope based on 10 CFR 54.4(a)(3). Therefore, the staff concern described in CI 2.3.3.12-1 is resolved.

CI 2.3.3.13a-1

The LRA states that the augmented off-gas system is within the scope of license renewal based on requirements of 10 CFR 54.4(a)(2) because of the potential for physical interaction with safety-related components described in LRA Table 2.3.3.13-A. The determination of whether a component meets the requirements of 10 CFR 54.4(a)(2) for physical interactions is based on where it is located in a building and its proximity to safety-related equipment or where a structural/seismic boundary exists. This information is not provided on license renewal drawings nor was a detailed description provided in the LRA. Consequently, any omission of augmented off-gas components subject to an AMR cannot be determined. The staff requested that the NRC Regional Inspection Team perform an inspection to ensure that the license renewal scope boundaries for these components meet the requirements of 10 CFR 54.4(a)(2) and all the components subject to an AMR are included in LRA Table 2.3.3-13-1.

In Inspection Report 05000271/2007006, Vermont Yankee Nuclear Power Station - NRC License Renewal Inspection Report, dated June 4, 2007, Attachment, Review of Safety Evaluation Report Confirmatory Items, the NRC Regional Inspection Team noted LRA Table 2.3.3.13-B states that the portion of the AOG system associated with the plant stack loop seal is subject to an AMR based on 10 CFR 54.4(a)(2) for physical interactions. Since the boundaries for the portion of the system as described in LRA Table 2.3.3.13-B were not well defined, in its letter dated July 30, 2007, the applicant amended the table to read "portion of the system inside the plant stack." The inspector walked down the remainder of the system and confirmed that no other portions of the system should have been included based on 10 CFR 54.4(a)(2).

Based on its review, the staff found the above response acceptable because the applicant amended LRA Table 2.3.3.13-B as appropriate and the NRC regional inspector walked down the remainder of the AOG system outside the plant stack and confirmed that no other portions of the system should have been included within scope based on 10 CFR 54.4(a)(2). Therefore, the staff concern described in CI 2.3.3.13a-1 is resolved.

CI 2.3.3.13e-1

The LRA states that the circulating water system is within the scope of license renewal based on the potential for physical interaction with safety-related components as required by 10 CFR 54.4(a)(2) and described in LRA Table 2.3.3.13-A. The applicant did not provide drawings highlighting in-scope components required by 10 CFR 54.4(a)(2), stating that the drawings would not provide significant additional information because they do not indicate proximity of components to safety-related equipment and do not identify structural/seismic boundaries. Without license renewal drawings and/or detailed description of the circulating water system, the omission of components subject to an AMR cannot be determined (see LRA Table 2.3.3-13-9). The staff requested that the NRC Regional Inspection Team perform an inspection to ensure that the license renewal scope boundaries for these components satisfy the requirements of 10 CFR 54.4(a)(2) and all the components subject to an AMR are included in LRA Table 2.3.3-13-9.

In Inspection Report 05000271/2007006, Vermont Yankee Nuclear Power Station - NRC License Renewal Inspection Report, dated June 4, 2007, Attachment, Review of Safety Evaluation Report Confirmatory Items, the NRC Regional Inspection Team stated that if any nonsafety-related portion of a fluid system is located within a building containing safety-related components, the components within the system are within the license renewal scope. Further, applicant's letter to the NRC dated July 3, 2007, LRA Amendment 27, Attachment 2 states that there are no nonsafety-related systems for which the applicant has not identified the nonsafety-related portions of systems which are attached to safety-related systems and required to be in the scope of license renewal in accordance with 10 CFR 54.4(a)(2). However, as a result of discussions with the staff during the Region I inspection (February 2007), the applicant determined that some safety-related SSCs in the VY turbine building required consideration for potential spatial impacts from nonsafety-related SSCs in accordance with 10 CFR 54.4(a)(2). Therefore, an expanded review for SSCs in the turbine building determined that additional components required an AMR. Those additional component types were added to LRA Table 2.3.3-13-9, as addressed in the applicant's letters to the staff dated July 30, 2007 and August 16, 2007.

Based on its review, the staff found the above response acceptable because the NRC Regional Inspection Team found that if any nonsafety-related portion of a fluid system is located within a building containing safety-related components, the components within the system are within the license renewal scope in accordance with 10 CFR 54.4(a)(2) but that there were spatial impact concerns from nonsafety-related SSCs in the turbine building. The additional component types have been added to LRA Table 2.3.3-13-9. Therefore, the staff concern regarding components of the CW system described in CI 2.3.3.13e-1 is resolved.

CI 2.3.3.13m-1

The LRA states that the reactor water clean up system is within the scope of license renewal in accordance with 10 CFR 54.4(a)(2) because of the potential for physical interaction with safety-related components as described in LRA Table 2.3.3.13-A. The determination of whether a component meets the requirements of 10 CFR 54.4(a)(2) for physical interactions is based on where it is located in a building and its proximity to safety-related equipment or where a structural/seismic boundary exists. This information is not provided on license renewal drawings nor was a detailed description provided in the LRA. Consequently, any omission of the reactor water clean up components subject to an AMR cannot be determined. The staff requested that the NRC Regional Inspection Team perform an inspection to ensure that the license renewal scope boundaries for these components satisfy the requirements of 10 CFR 54.4(a)(2) and all the components subject to an AMR are included in LRA Table 2.3.3-13-36.

In Inspection Report 05000271/2007006, Vermont Yankee Nuclear Power Station - NRC License Renewal Inspection Report, dated June 4, 2007, Attachment, Review of Safety Evaluation Report Confirmatory Items, the NRC Regional Inspection Team stated that if any nonsafety-related portion of a fluid system is located within a building containing safety-related components, the components within the system are within the license renewal scope. Further, the applicant's letter to the NRC dated July 3, 2007, LRA Amendment 27, Attachment 2 states that there are no nonsafety-related systems for which the applicant has not identified the nonsafety-related portions of systems which are attached to safety-related systems and required to be in the scope of license renewal in accordance with 10 CFR 54.4(a)(2). The applicant also stated that there were no additional components that should be within scope based on 10 CFR 54.4(a) as identified during the NRC Regional Inspection and subsequent applicant reviews.

Based on its review, the staff found the above response acceptable because the NRC Regional Inspection Team found that if any nonsafety-related portion of a fluid system is located within a building containing safety-related components, the components within the system are within the license renewal scope in accordance with 10 CFR 54.4(a)(2) and that there were no additional components identified that should be in-scope based on 10 CFR 54.4(a). Therefore, the staff concern regarding the components of the RWCU system described in CI 2.3.3.13m-1 is resolved.

1.7 Summary of Proposed License Conditions

Following the staff's review of the LRA, including subsequent information and clarifications provided by the applicant, the staff identified four proposed license conditions.

The first license condition requires the applicant to include the UFSAR supplement required by 10 CFR 54.21(d) in the next UFSAR update, as required by 10 CFR 50.71(e), following the issuance of the renewed license.

The second license condition requires future activities identified in the UFSAR supplement to be completed prior to the period of extended operation.

The third license condition requires the implementation of the most recent staff-approved version of the Boiling Water Reactor Vessels and Internals Project (BWRVIP) Integrated Surveillance Program (ISP) as the method to demonstrate compliance with the requirements of 10 CFR Part 50, Appendix H. Any changes to the BWRVIP ISP capsule withdrawal schedule must be submitted for NRC staff review and approval. Any changes to the BWRVIP ISP capsule withdrawal schedule which affects the time of withdrawal of any surveillance capsules must be incorporated into the licensing basis. If any surveillance capsules are removed without the intent to test them, these capsules must be stored in a manner which maintains them in a condition which would support re-insertion into the reactor pressure vessel, if necessary.

The fourth license condition requires that the licensee perform and submit to the NRC for review and approval, a ASME Code analysis for the reactor recirculation outlet nozzle and the core spray nozzle at least two years prior to the period of extended operation. These analyses should be documented in the FSAR as the analysis-of-record for these two nozzles.

SECTION 2

STRUCTURES AND COMPONENTS SUBJECT TO AGING MANAGEMENT REVIEW

2.1 Scoping and Screening Methodology

2.1.1 Introduction

Title 10, Section 54.21, of the *Code of Federal Regulations* (CFR), "Contents of Application Technical Information" (10 CFR 54.21), requires for each license renewal application (LRA) an integrated plant assessment (IPA) listing structures and components (SCs) subject to an aging management review (AMR) from all of the systems, structures, and components (SSCs) within the scope of license renewal.

LRA Section 2.1, "Scoping and Screening Methodology," describes the methodology for identifying SSCs at the Vermont Yankee Nuclear Power Station (VYNPS) within the scope of license renewal and SCs subject to an AMR. The staff of the United States (US) Nuclear Regulatory Commission (NRC) (the staff) reviewed the Entergy Nuclear Operations, Inc. (ENO or the applicant) scoping and screening methodology to determine whether it meets the scoping requirements of 10 CFR 54.4(a) and the screening requirements of 10 CFR 54.21.

In developing the scoping and screening methodology for the LRA, the applicant considered the requirements of 10 CFR Part 54, "Requirements for Renewal of Operating Licenses for Nuclear Power Plants" (the Rule), statements of consideration on the Rule, and the guidance of Nuclear Energy Institute (NEI) 95-10, Revision 6, "Industry Guideline for Implementing the Requirements of 10 CFR Part 54 - The License Renewal Rule," dated June 2005. The applicant also considered the correspondence between the staff, other applicants, and the NEI.

2.1.2 Summary of Technical Information in the Application

LRA Sections 2 and 3 state the technical information required by 10 CFR 54.4 and 54.21(a). LRA Section 2.1 describes the process for identifying SSCs meeting the license renewal scoping criteria of 10 CFR 54.4(a) and the process for identifying SCs subject to an AMR as required by 10 CFR 54.21(a)(1). The applicant provided the results of the process for identifying such SCs in the following LRA sections:

- Section 2.2, "Plant Level Scoping Results"
- Section 2.3, "Scoping and Screening Results: Mechanical Systems"

- Section 2.4, "Scoping and Screening Results: Structures"
- Section 2.5, "Scoping and Screening Results: Electrical and Instrumentation and Control Systems"

LRA Section 3, "Aging Management Review Results," states the applicant's aging management results in the following LRA sections:

- Section 3.1, "Reactor Vessel, Internals and Reactor Coolant System"
- Section 3.2, "Engineered Safety Features Systems"
- Section 3.3, "Auxiliary Systems"
- Section 3.4, "Steam and Power Conversion Systems"
- Section 3.5, "Structures and Component Supports"
- Section 3.6, "Electrical and Instrumentation and Controls"

LRA Section 4, "Time-Limited Aging Analyses," states the applicant's evaluation of time-limited aging analyses.

2.1.3 Scoping and Screening Program Review

The staff evaluated the LRA scoping and screening methodology in accordance with the guidance in Section 2.1, NUREG-1800, "Standard Review Plan for Review of License Renewal Applications for Nuclear Power Plants," Revision 1, (SRP-LR), and the Nuclear Energy Institute (NEI) 95-10, "Industry Guidelines for Implementing the Requirements of 10 CFR Part 54 - The License Renewal Rule," Revision 6, (NEI 95-10). The following regulations form the basis for the acceptance criteria for the scoping and screening methodology review:

- 10 CFR 54.4(a) as to identification of plant SSCs within the scope of the Rule
- 10 CFR 54.4(b) as to identification of the intended functions of plant systems and structures within the scope of the Rule
- 10 CFR 54.21(a)(1) and 10 CFR 54.21(a)(2) as to the methods utilized by the applicant to identify plant SCs subject to an AMR

With the guidance of the corresponding SRP-LR sections, the staff reviewed, as part of the applicant's scoping and screening methodology, the activities described in the following LRA sections:

- Section 2.1 to ensure that the applicant described a process for identifying SSCs within the scope of license renewal in accordance with 10 CFR 54.4(a)
- Section 2.2 to ensure that the applicant described a process for identifying SCs subject to an AMR in accordance with 10 CFR 54.21(a)(1) and 10 CFR 54.21(a)(2)

The staff conducted a scoping and screening methodology audit at VYNPS in Vernon, Vermont during the week of April 24-28, 2006. The audit focused on whether the applicant had developed and implemented adequate guidance for the scoping and screening of SSCs by the methodologies in the LRA and the requirements of the Rule. The staff reviewed implementation of the project level guidelines and topical reports describing the applicant's scoping and screening methodology. The staff discussed with the applicant details of the implementation and control of the license renewal program and reviewed administrative control documentation and selected design documentation used by the applicant during the scoping and screening process. The staff reviewed the applicant's processes for quality assurance (QA) for development of the LRA. The staff reviewed the quality attributes of the applicant's aging management program (AMP) activities described in LRA Appendix A, "Updated Final Safety Analysis Report Supplement," and LRA Appendix B, "Aging Management Programs and Activities" and the LRA training and qualification development team. The staff reviewed scoping and screening results reports for the core spray (CS) system and intake structure for the applicant's appropriate implementation of the methodology outlined in the administrative controls and for results consistent with the current licensing basis (CLB) documentation.

2.1.3.1 Implementation Procedures and Documentation Sources for Scoping and Screening

The staff reviewed the applicant's scoping and screening implementation procedures as documented in the audit report dated August 10, 2006 to verify whether the process for identifying SCs subject to an AMR was consistent with the LRA and the SRP-LR. Additionally, the staff reviewed the scope of CLB documentation sources and the applicant's process for appropriate consideration of CLB commitments and for adequate implementation of the procedural guidance during the scoping and screening process.

2.1.3.1.1 Summary of Technical Information in the Application

In LRA Section 2.1, the applicant addressed the following information sources for the license renewal scoping and screening process:

- System and Topical Design Basis Documents (DBDs)
- VYNPS Enterprise Maintenance, Planning, and Control (EMPAC) Component Database
- Updated Final Safety Analysis Report (UFSAR)
- Appendix R Safe Shutdown Capability Assessment
- Fire Hazards Analysis Report
- Safe Shutdown Capability Assessment
- Technical Specifications
- Maintenance Rule Scoping Basis Documents
- Safety Classification Documents
- Plant Layout Drawings

The applicant stated that it used this information to identify the functions performed by plant systems and structures. It then compared these functions to the scoping criteria in 10 CFR 54.4(a)(1-3) to determine whether the associated plant system or structure performed a license renewal intended function. It also used these sources to develop the list of SCs subject to an AMR.

The license renewal boundary drawings (LRBDs) show the systems within the scope of license renewal highlighted in color.

2.1.3.1.2 Staff Evaluation

<u>Scoping and Screening Implementation Procedures</u>. The staff reviewed the following scoping and screening methodology implementation procedures:

The staff reviewed the applicant's scoping and screening methodology implementation procedures, including license renewal project guidelines (LRPGs), license renewal project documents/reports (LRPDs), AMR reports (e.g., AMRMs - mechanical, AMREs- electrical, and AMRCs - structural), as documented in the audit report, to ensure the guidance was consistent with the requirements of the Rule, NUREG-1800, "Standard Review Plan for Review of License Renewal Applications for Nuclear Power Plants," Revision 1, (SRP-LR), and the Nuclear Energy Institute (NEI) 95-10, "Industry Guidelines for Implementing the Requirements of 10 CFR Part 54 - The License Renewal Rule," Revision 6, (NEI 95-10).

The staff found the overall process for implementing 10 CFR Part 54 requirements included in the LRPGs, LRPDs, and AMRs was consistent with the Rule and industry guidance. The staff found guidance for identifying plant SSCs within the scope of the Rule, including guidelines for identifying SC component types within the scope of license renewal subject to an AMR, in the LRA, including in the implementation of NRC staff positions documented in NUREG-1800, and the information in requests for additional information (RAI) responses dated July 10, 2006. The review of these procedures focused on the consistency of the detailed procedural guidance with information in the LRA reflecting implementation of staff positions in the SRP-LR and interim staff guidance documents.

After reviewing the LRA and supporting documentation, the staff finds LRA Section 2.1 consistent with the scoping and screening methodology instructions. The applicant's methodology has sufficiently detailed guidance for the scoping and screening implementation process followed in the LRA.

<u>Sources of Current Licensing Basis Information</u>. For VYNPS, system safety functions are stated in safety classification documents, the Maintenance Rule SSC basis documents for each system, and in design basis documents for systems for which DBDs were written. The staff considered the safety objectives in the UFSAR system descriptions and identified objectives meeting the safety-related Criterion of 10 CFR 54.4(a)(1) as system intended functions.

The staff reviewed the scope and depth of the applicant's CLB information to verify whether the applicant's methodology had identified all SSCs within the scope of license renewal as well as component types requiring AMRs. As defined in 10 CFR 54.3(a), the CLB applies NRC requirements, written licensee commitments for compliance with, and operation within, applicable NRC requirements, and plant-specific design bases docketed and in effect. The CLB includes NRC regulations, orders, license conditions, exemptions, technical specifications, design-basis information in the most recent UFSAR, and licensee commitments in docketed correspondence like licensee responses to NRC bulletins, generic letters, and enforcement actions as well as commitments in NRC safety evaluations or licensee event reports.

During the audit, the staff reviewed the applicant's information sources and samples of such information, including the UFSAR, DBDs, controlled plant reference drawings, LRBDs, and Maintenance Rule information. In addition, the applicant's license renewal process identified additional potential sources of plant information pertinent to the scoping and screening process, including, licensing correspondence, the Fire Hazards Analysis, safety evaluations, and design documentation such as engineering calculations and design specifications. Additionally, the applicant supplemented the review by using an electronic database developed during the plant FSAR accuracy verification project conducted between 1998 and 2000. The database contained approximately 10,000 documents including all correspondence in the public document room. The searchable database was available for query during the review of the CLB information in support of LRA development. The staff confirmed that the applicant's detailed license renewal program guidelines required use of the CLB source information developing scoping evaluations.

The VYNPS component database is the applicant's primary repository for component safety classification information. During the audit, the staff reviewed the applicant's administrative controls for VYNPS component database safety classification data. These controls are described and implementation is governed by plant administrative procedures. Based on a review of the administrative controls, and a sample of the VYNPS component database component safety classifications, the NRC staff concluded that the applicant had established adequate measures to control the integrity and reliability of VYNPS component database safety classification data, and therefore, the staff concluded that the VYNPS component database provided a sufficiently controlled source of component data to support scoping and screening evaluations.

During the staff's review of the applicant's CLB evaluation process, the applicant provided the staff with a discussion regarding the incorporation of updates to the CLB and the process used to ensure those updates are adequately incorporated into the license renewal process. The staff determined that LRA Section 2.1 provided a description of the CLB and related documents used during the scoping and screening process that is consistent with the guidance contained in NUREG-1800. In addition, the staff reviewed technical reports utilized to support identification of SSCs relied upon to demonstrate compliance with the safety-related criteria, nonsafety-related criteria, as well as the five regulated events referenced in 10 CFR 54.4(a)(1-3). The applicants license renewal program guidelines provided a comprehensive listing of documents used to support scoping and screening evaluations. The staff found these design documentation sources to be useful for ensuring that the initial scope of SSCs identified by the applicant was consistent with the plant's CLB.

2.1.3.1.3 Conclusion

Based on its review of LRA Section 2.1, the detailed scoping and screening implementation procedures, and the results from the scoping and screening audit, the staff concludes that the applicant's scoping and screening methodology considers CLB information consistently with SRP-LR and NEI 95-10 guidance and, therefore, is acceptable.

2.1.3.2 Quality Controls Applied to LRA Development

2.1.3.2.1 Staff Evaluation

The staff reviewed the quality controls used by the applicant to ensure that scoping and screening methodologies described in the LRA were adequately implemented. Although the applicant did not develop the LRA in accordance with a 10 CFR 50, Appendix B, QA program, the applicant utilized the following QA processes during the LRA development:

- Implementation of the scoping and screening methodology was governed by written procedures.

- The applicant reviewed previous LRA NRC requests for additional information to ensure that applicable issues were addressed in the LRA.

- The LRA was reviewed by the Offsite and Onsite Safety Review Committees prior to submittal to the NRC.

- The applicant performed an industry peer review of the LRA.

- The applicant's QA organization performed an independent review of the LRA. The purpose of this review was to ensure that the technical information used to develop the LRA was updated and approved in accordance with the station's QA program, and that industry peer and Offsite and Onsite Safety Review Committee issues were resolved and associated corrective actions implemented.

2.1.3.2.2 Conclusion

Based on its review of pertinent LRA development guidance, discussion with the applicant's license renewal personnel, and review of the quality audit reports, the staff concludes that these QA activities add assurance that LRA development activities have been performed in accordance with the scoping and screening methodologies described in the LRA.

2.1.3.3 Training

2.1.3.3.1 Staff Evaluation

The staff reviewed the applicant's training process for consistent and appropriate guidelines and methodology for the scoping and screening activities and to ensure the guidelines and methodology were performed in a consistent and appropriate manner.

The LRPGs provided the guidance and requirements for the training of the license renewal project and site personnel. The training consisted of a combination of reading and attending training sessions. The LRPGs specified the level of training which was required for the various groups participating in the development of the LRA and began with initial training, documented on a qualification card. The training was required for both the license renewal project personnel who prepared the application and for the site personnel who reviewed the application. In addition, license renewal refresher training was provided for the license renewal project and site personnel participating in the review. Refresher training included information on the license renewal process and information specific to the site. License renewal project and site personnel were required to review applicable license renewal regulations, NEI 95-10 and associated procedures. The applicant developed periodic production meetings in which the license renewal project personnel shared their knowledge and experience of a given subject with each other.

The NRC staff reviewed completed qualification and training records of several of the applicant's license renewal project personnel and also reviewed completed check lists. The staff found these records adequately documented the required training for the license renewal project personnel. Additionally, based on discussions with the applicant's license renewal project personnel during the audit, the NRC staff confirmed that the applicant's license renewal project personnel were knowledgeable on the license renewal process requirements and the specific technical issues within their areas of responsibility.

On the basis of discussions with the applicant's license renewal project personnel responsible for the scoping and screening process, and a review of selected design documentation in support of the process, the NRC staff concluded that the applicant's license renewal project personnel understood the requirements of and adequately implemented the scoping and screening methodology established in the applicant's renewal application. The staff did not identify any concerns regarding the training of the applicant's license renewal project or site personnel.

2.1.3.3.2 Conclusion

Based on discussions with the applicant's license renewal personnel responsible for the scoping and screening process and review of selected documentation supporting the process, the staff concludes that the applicant's technical personnel understood the requirements and adequately implemented the scoping and screening methodology documented in the LRA. The staff concludes that the license renewal personnel were adequately trained and qualified for license renewal activities.

2.1.3.4 Conclusion of Scoping and Screening Program Review

Based on its review of LRA Section 2.1, review of the applicant's detailed scoping and screening implementation procedures, discussions with the applicant's LRA personnel, and review of the scoping and screening audit results, the staff concludes that the applicant's scoping and screening program is consistent with SRP-LR guidance and, therefore, acceptable.

2.1.4 Plant Systems, Structures, and Components Scoping Methodology

LRA Section 2.1, describes the methodology for scoping SSCs as required by 10 CFR 54.4(a) and the plant scoping process for systems and structures. Specifically, the scoping process consisted of developing a list of plant systems and structures and identifying their intended functions. Intended functions are those functions that are the basis for including a system or structure within the scope of license renewal as defined in 10 CFR 54.4(b) and are identified by comparing the system or structure function with the criteria in 10 CFR 54.4(a). The systems list was developed from the VYNPS component database and the structures list from a review of plant layout drawings and structure-specific system codes in the VYNPS component database.

Finally, the applicant evaluated the components in the systems and structures that were in-scope of license renewal. The in-scope system boundary of SSCs subject to an AMR is depicted on the license renewal drawings. The applicant's scoping methodology, as described in the LRA, is discussed in the sections below.

2.1.4.1 Application of the Scoping Criteria in 10 CFR 54.4(a)(1)

2.1.4.1.1 Summary of Technical Information in the Application

In LRA Section 2.1.1.1, "Application of Safety-Related Scoping Criteria," the applicant described the scoping methodology required by 10 CFR 54 as it relates to safety-related criteria in accordance with 10 CFR 54.4(a)(1). With respect to the safety-related criteria, the applicant stated that at VYNPS system safety functions are identified in safety classification documents, the maintenance rule SSC basis documents for each system, and in design basis documents (DBDs) for those systems for which a DBD was written. SSCs that are identified as safety-related in the UFSAR, in DBDs, or in the CRL were classified as satisfying criteria of 10 CFR 54.4(a)(1) and included within the scope of license renewal. The review also confirmed that all plant conditions, including conditions of normal operation, abnormal operational transients, design basis accidents, internal and external events, and natural phenomena for which the plant must be designed, were considered for license renewal scoping in accordance with 10 CFR 54.4(a)(1) criteria.

The VYNPS CLB definition of safety-related SSCs is not identical to the definition provided in the Rule. As a result, the applicant performed an evaluation of the differences between its CLB definition of safety-related and the Rule definition.

2.1.4.1.2 Staff Evaluation

Pursuant to 10 CFR 54.4(a)(1), the applicant must consider all safety-related SSCs relied upon to remain functional during and following a design basis event (DBE) to ensure (a) the integrity of the reactor coolant pressure boundary, (b) the ability to shut down the reactor and maintain it in a safe shutdown condition, or (c) the ability to prevent or mitigate the consequences of accidents that could cause offsite exposures comparable to those of 10 CFR 50.34(a)(1), 10 CFR 50.67(b)(2), or 10 CFR 100.11.

As to identification of DBEs, SRP-LR Section 2.1.3 states:

> The set of DBEs as defined in the Rule is not limited to Chapter 15 (or equivalent) of the UFSAR. Examples of DBEs that may not be described in this chapter include external events, such as floods, storms, earthquakes, tornadoes, or hurricanes, and internal events, such as a high-energy line break. Information regarding DBEs as defined in 10 CFR 50.49(b)(1) may be found in any chapter of the facility UFSAR, the Commission's regulations, NRC orders, exemptions, or license conditions within the CLB. These sources should also be reviewed to identify SSCs relied upon to remain functional during and following DBEs (as required by 10 CFR 50.49(b)(1)) to ensure the functions required by 10 CFR 54.4(a)(1).

The staff's review of LRA Section 2.1 of VYNPS identified areas in which additional information was necessary to complete the review of the applicant's scoping and screening methodology. The applicant responded to the staff's RAIs as discussed below.

During the scoping and screening methodology audit, the staff questioned how non-accident DBEs, particularly DBEs that may not be described in the UFSAR, were considered during scoping. The staff noted that limiting the review of DBEs to those described in the UFSAR accident analysis could result in omission of safety-related functions described in the CLB and requested the applicant provide a list of all DBEs that were evaluated as part of the license renewal review. However, during the audit, the staff was unable to identify such a list. Therefore, in RAI 2.1-1, dated July 10, 2006, the staff requested that the applicant provide: a) a list of DBEs evaluated as part of the license renewal scoping process, b) describe the methodology used to ensure that all DBEs (including conditions of normal operation, anticipated operational occurrences, design-basis accidents, external events, and natural phenomena) were addressed during license renewal scoping evaluation, and c) a list of the documentation sources reviewed to ensure that all DBEs were identified.

In its response, by letter dated August 10, 2006, the applicant described the DBEs evaluated during the license renewal effort and described the methodology used to ensure that all DBEs were addressed during license renewal scoping. Specifically, the applicant identified abnormal operational transients, design-basis accidents, events for which the alternate cooling system (ACS) is credited (i.e., loss of the Vernon Pond and flooding or fire in the service water (SW) intake structure), and additional DBEs such as external and internal flooding, earthquakes, tornadoes and natural phenomena as constituting the DBEs for the Vermont Yankee plant.

In addition, the applicant described two basic means of ensuring that all of the plant DBEs were addressed during the license renewal scoping process. These include: (1) reviewing the UFSAR and DBDs (i.e., for external and internal events and safety analyses) directly for the identification of the DBEs and subsequently for the identification of the SSCs credited for each event, and (2) reviewing and evaluating the safety classification of systems and components as governed by the plant safety classification process. This process ensures that site-specific procedures, design basis information, regulatory commitments, and regulatory guidance are considered during the classification process. The VYNPS safety classification process identifies those SSCs which are credited for performance of the intended safety functions in accordance with 10 CFR 54.4(a)(1).

The NRC staff reviewed a sample of the DBDs identified as sources of this information. The staff found the DBDs to contain a detailed evaluation of events, and included appropriate CLB documentation references to support the review and a resultant matrix of systems and structures relied upon to remain functional during and following these DBEs. The staff concluded that the applicant considered DBEs consistent with the guidance contained in NUREG-1800.

The staff reviewed the additional information provided by the applicant and, on the basis of providing (1) a detailed listing of the DBEs for the plant; (2) a description of the design and configuration control processes used to identify the SSCs credited for DBE mitigation; and (3) a description of the processes and sources of DBE information used to perform the scoping evaluation consistent with the requirements of 10 CFR 54.4(a)(1), the staff found that the applicant has adequately addressed the staff's RAI. Therefore, the staff's concern described in RAI 2.1-1 is resolved.

The applicant performed scoping of SSCs for the 10 CFR 54.4(a)(1) criterion in accordance with the LRPGs which provided guidance for the preparation, review, verification, and approval of the scoping evaluations to assure the adequacy of the results of the scoping process. The staff reviewed these guidance documents governing the applicant's evaluation of safety-related SSCs, and sampled the applicant's scoping results reports to ensure the methodology was implemented in accordance with those written instructions. In addition, the staff discussed the methodology and results with the applicant's technical personnel who were responsible for these evaluations.

The staff reviewed a sample of the license renewal scoping results for the CSS and the Intake Structure to provide additional assurance that the applicant adequately implemented their scoping methodology with respect to 10 CFR 54.4(a)(1). The staff confirmed that the scoping results for each of the sampled systems were developed consistent with the methodology, the SSCs credited for performing intended functions were identified, and the basis for the results as well as the intended functions were adequately described. The staff confirmed that the applicant had identified and used pertinent engineering and licensing information to identify the SSCs required to be in-scope in accordance with the 10 CFR 54.4(a)(1) criteria.

To document the identification of SSCs in-scope in accordance with the 10 CFR 54.4(a) criteria, the applicant developed a scoping report which contained detailed design description information about each plant system and structure and the relevant functions of those systems and structures. A list of safety-related SCs was initially identified by using the existing components list in the VYNPS component database. The VYNPS component database safety-classification field was reviewed to ensure that any system or structure that has a component identified as safety-related was considered for inclusion into the scope of the license renewal project. For VYNPS, component safety classification fields SC1 - SC3 corresponded to the 10 CFR 54.4(a)(1) criteria. Additionally, the SC1 database safety-classification and associated plant system drawings provided a starting point for identifying specific components which were required to meet the 10 CFR 54.4(a)(1) criteria.

During the audit, the applicant described the process used to evaluate components classified as safety-related that did not perform a safety-related intended function. As part of the process, the applicant stated that the safety-classification of several components were reevaluated in order to reconcile differences between scoping determinations and facility database information or CLB information. Those components that were identified as safety-related that did not perform an intended function were explicitly evaluated and described in the LRPD's and the rationale for their exclusion from scope of the license renewal was documented. For instances where components identified as safety-related in the VYNPS component database did not perform any safety-related functions, the applicant identified these components and performed additional evaluations to confirm that the component did not perform or were not credited in the CLB for any specific safety-related functions. Examples included the reactor water cleanup (RWCU) system and the augmented off-gas (AOG) system.

The staff reviewed the safety classification criteria used to determine the safety classification to verify consistency between the VYNPS CLB definition and the Rule definition in 10 CFR 54.4(a). In addition, the staff reviewed the applicant's evaluation of the differences between the Rule definition and the site-specific definition of safety-related to ensure all potential SSCs meeting the requirements of 10 CFR 54.4(a)(1) were adequately addressed. The applicant documented this evaluation in the LRA and LRPDs. As part of the license renewal development activities, the applicant stated that the site-specific definition for safety-related was nearly identical to the Rule definition with the following exception:

> The CLB definition regarding potential offsite exposure limits refers to 10 CFR 50.67 whereas the Rule also references comparable guidelines in 10 CFR 50.34(a)(1), 10 CFR 50.67(b)(2), and 10 CFR Part 100 respectively.

During the audit, the staff reviewed the applicant's evaluation of the Rule and VY CLB definitions pertaining to 10 CFR 54.4(a)(1). Based on this review, the staff confirmed that 10 CFR 50.34(a)(1)(ii) is not applicable to VYNPS as it concerns applicants for a construction permit who apply on or after January 10, 1997. In addition, the staff has amended the VYNPS operating license to allow use of an alternative source term for accident analyses in accordance with 10 CFR 50.67. The change to 10 CFR 50.67 dose limits does not affect the VYNPS safety classification definition. The accident analyses with the alternative source term credits additional functions for the standby liquid control (SLC) and residual heat removal (RHR) systems: (1) the SLC system is credited with maintaining pH in the torus to prevent re-evolution of iodine, and (2) the drywell spray function of the RHR system is credited with particulate removal. The staff confirmed that these intended functions were included in the scoping evaluation.

During the audit, the staff also confirmed that any SSCs specifically credited for the 10 CFR 50.67(b) leakage pathway, were identified and included in-scope. For VYNPS, the main condenser and main steam (MS) bypass leakage pathway are credited for 10 CFR 50.67(b) leakage pathway and meet the 10 CFR 54.4(a)(1)(iii) criterion for inclusion in-scope. The staff confirmed that these pertinent SSCs were appropriately identified and placed in-scope. Since the specific SSCs were classified as nonsafety-related in the plant component database, they were placed in-scope in accordance with 10 CFR 54.4(a)(2) for nonsafety-related potentially affecting a safety-related functions.

The staff reviewed the evaluation and discussed the results of the evaluation with the applicant's license renewal team members. The staff determined that the differences between the VYNPS safety-related definition and the Rule definition were adequately identified and evaluated. These differences did not result in any additional components being considered safety-related beyond those identified in the VYNPS CLB.

2.1.4.1.3 Conclusion

Based on this sample review, discussions with the applicant, and review of the applicant's scoping process, the staff finds that the applicant's methodology for identifying systems and structures meets 10 CFR 54.4(a)(1) scoping criteria and, therefore, is acceptable.

2.1.4.2 Application of the Scoping Criteria in 10 CFR 54.4(a)(2)

2.1.4.2.1 Summary of Technical Information in the Application

In LRA Section 2.1.1.2, "Application of Criterion for Nonsafety-Related SSCs Whose Failure Could Prevent the Accomplishment of Safety Functions," and Section 2.3.3.13, "Miscellaneous Systems in-Scope for (a)(2)," the applicant described the scoping methodology as it related to the nonsafety-related criteria in accordance with 10 CFR 54.4(a)(2). The applicant evaluated the SSCs that met 10 CFR 54.4(a)(2) using three categories:

(1)   Nonsafety-Related SSCs Required to Perform a Function that Supports a Safety-Related SSC

The SSCs required to perform a function in support of safety-related components were classified as safety-related and included in the scope of license renewal in accordance with 10 CFR 54.4(a)(1). The applicant reviewed engineering and licensing documents (UFSAR, Maintenance Rule scoping documents, and DBDs) to identify exceptions which were included within the scope of license renewal in accordance with 10 CFR 54.4(a)(2).

(2)   Nonsafety-Related SSCs Connected to Safety-Related SSCs

The applicant identified certain nonsafety-related components and piping outside of the safety-class pressure boundary which must be structurally sound in order to maintain the pressure boundary integrity of safety-related piping. These components perform a structural support function.

For piping in this structural boundary, pressure integrity is not required (except when required for spatial interaction between nonsafety-related and safety-related SSCs); however, piping within the safety class pressure boundary depends on the structural boundary piping and supports in order for the system to fulfill its safety function. For VYNPS, the "structural boundary" is defined as the portion of a piping system outside the safety class pressure boundary, yet relied upon to provide structural support for the pressure boundary. The structural boundary is often shown on piping isometric drawings and was considered synonymous with the first seismic or equivalent anchor. Nonsafety-related piping systems connected to safety-related systems were included up to the structural boundary or to a point that includes an adequate portion of the nonsafety-related piping run to conservatively include the first seismic or equivalent anchor. An equivalent anchor was a combination of hardware or structures that together are equivalent to a seismic anchor. A seismic anchor was defined as hardware or structures that, as required by the analysis, physically restrain forces and moments in three orthogonal directions. The physical arrangement as analyzed insures that the stresses that are developed in the safety-related piping and supports are within the applicable piping and structural code acceptance limits. If isometric drawings were not readily available to identify the structural boundary, connected lines were included to a point beyond the safety-related/nonsafety-related interface, such as a base-mounted component, flexible connection, or the end of a piping run (such as a drain line). The LRA stated that the approach was consistent with the guidance in NEI 95-10, Appendix F.

(3)   Nonsafety-related SSCs with a Potential for Spatial Interaction with Safety-Related SSCs

The applicant considered physical impact, and fluid leakage, spray or flooding when evaluating the potential for spatial interaction between nonsafety-related systems and safety-related SSCs. The applicant used a spaces approach for scoping of nonsafety-related systems with potential spatial interaction with safety-related SSCs. The spaces approach focused on the interaction between nonsafety-related and safety-related SSCs that are located in the same space. A "space" was defined as a room or cubicle that is separated from other spaces by substantial objects (such as wall, floors, and ceilings). The space was defined such that any potential interaction between nonsafety-related and safety-related SSCs is limited to the space.

Physical Impact or Flooding

The applicant evaluated missiles which could be generated from internal or external events such as failure of rotating equipment. Inherent nonsafety-related features that protect safety-related equipment from missiles; overhead-handling systems whose structural failure could result in damage to any system that could prevent the accomplishment of a safety function; and walls, curbs, dikes, doors, etc, that provide flood barriers to safety-related SSCs meet the criteria of 10 CFR 54.4(a)(2). Nonsafety-related equipment that was determined to have a possible impact on safety-related SSCs were included within the scope of license renewal.

The applicant evaluated nonsafety-related portions of high-energy lines, including review of the UFSAR and relevant topical design basis document. The applicant's high-energy systems were evaluated to ensure identification of components that are part of nonsafety-related high-energy lines that can effect safety-related equipment. If the applicant's high-energy line break (HELB) analysis assumed that an nonsafety-related piping system did not fail or assumed failure only at specific locations, then that piping system (piping, equipment and supports) is included within the scope of license renewal.

Fluid Leakage or Spray

The applicant evaluated moderate and low energy systems which have the potential for spatial interactions of spray and leakage. Nonsafety-related systems and nonsafety-related portions of safety-related systems with the potential for spray or leakage that could prevent safety-related SSCs from performing their required safety function were considered in the scope of license renewal. In addition, the nonsafety-related supports for nonsafety-related piping systems with a potential for spatial interaction with safety-related SSCs were included in the scope of license renewal.

The applicant determined that operating experience indicated that nonsafety-related components containing only air or gas have experienced no failures due to aging that could impact the ability of safety-related equipment to perform required safety functions. There are no effects of aging requiring management for these components when the environment is a dry gas. Systems containing only air or gas were not included in the scope of license renewal.

Protective features, such as whip restraints, spray shields, supports, missile or flood barriers, (which can be applicable preventing physical impact and fluid leakage, spray, or flooding) were installed to protect safety-related SSCs against spatial interaction with nonsafety-related SSCs. Such protective features credited in the plant design were included within the scope of license renewal.

2.1.4.2.2 Staff Evaluation

Pursuant to 10 CFR 54(a)(2), the applicant must consider all nonsafety-related SSCs, the failure of which could prevent satisfactory performance of safety-related SSCs relied upon to remain functional during and following a DBE to ensure (a) the integrity of the reactor coolant pressure boundary, (b) the ability to shut down the reactor and maintain it in a safe shutdown condition, or (c) the ability to prevent or mitigate the consequences of accidents that could cause offsite exposures comparable to those of 10 CFR 50.34(a)(1), 10 CFR 50.67(b)(2), or 10 CFR 100.11, as applicable.

NRC Regulatory Guide (RG) 1.188, Revision 1, "Standard Format and Content for Applications to Renew Nuclear Power Plant Operating Licenses," dated September 2005, endorses the use of NEI 95-10, Revision 6, for methods the staff considers acceptable for compliance with 10 CFR Part 54 in preparing LRAs. NEI 95-10, Revision 6, addresses the staff positions on 10 CFR 54.4(a)(2) scoping criteria, nonsafety-related SSCs typically identified in the CLB, consideration of missiles, cranes, flooding, high-energy line breaks, nonsafety-related SSCs connected to safety-related SSCs, nonsafety-related SSCs in proximity of safety-related SSCs, and the mitigative and preventive options in nonsafety-related and safety-related SSCs interactions.

The staff states that applicants should not consider hypothetical failures but rather base their evaluation on the plant's CLB, engineering judgement and analyses, and relevant operating experience, describing operating experience as all documented plant-specific and industry-wide experience useful in determining the plausibility of a failure. Documentation would include NRC generic communications and event reports, plant-specific condition reports, such industry reports as safety operational event reports, and engineering evaluations.

The staff reviewed LRA Section 2.1.1.2, "Application of Criterion for Nonsafety-Related SSCs Whose Failure Could Prevent the Accomplishment of Safety Functions," and Section 2.3.3.13, "Miscellaneous Systems in-Scope for (a)(2)." The applicant described the scoping methodology as it related to the nonsafety-related criteria in accordance with 10 CFR 54.4(a)(2).

The applicant evaluated 10 CFR 54.4(a)(2) SSCs with the three categories from the NRC guidance to the industry on identification and treatment of such SSCs:

Nonsafety-Related SSCs Required to Perform Functions that Support a Safety-Related SSCs

Nonsafety-related SSCs required to perform a function in order to support a safety-related function had been previously classified as safety-related and were identified as such in the equipment data base. Therefore the nonsafety-related SSCs required to perform a function to support a safety-related function had been included in the scope of license renewal as safety-related as required by 10 CFR 54.4(a)(1). This evaluating criteria was discussed in the

2-14

applicant's 10 CFR 54.4(a)(2) project report. The single exception to this approach was the main condenser and main steam isolation valve (MSIV) leakage pathway which was classified as an nonsafety-related system and was required to perform a function to support a safety-related function. This system was included in the scope of license renewal in accordance with 10 CFR 54.4(a)(2). The staff found that the applicant implemented an acceptable method for scoping of nonsafety-related systems that perform a function that supports a safety-related intended function.

Nonsafety-Related SSCs Connected to Safety-Related SSCs

The applicant had previously performed an analysis to identify the nonsafety-related SSCs, outside of the safety-related pressure boundary, which were required to be structurally sound in order to maintain the integrity of the safety-related SSCs. This collection of nonsafety-related and safety-related SSCs was identified as the "structural boundary" and was typically identified on the plant isometric drawings. The applicant had included all nonsafety-related SSCs within the analyzed structural boundary in the scope of license renewal in accordance with 10 CFR 54.4(a)(2). The LRA states that if the structural boundary was not indicated on the applicable isometric drawings, the applicant had identified the portion of the nonsafety-related SSCs beyond the safety-related SSCs to the first equivalent anchor or seismic anchor and included this portion of the nonsafety-related SSCs within the scope of license renewal. The term equivalent anchor was defined in the LRA as a combination of hardware or structures that together are equivalent to a seismic anchor (a seismic anchor was defined as hardware or structures that, as required by analysis, physically restrain forces and moments in three orthogonal directions). The LRA also indicated that if the structural boundary could not be identified for the applicable nonsafety-related/safety-related interface, the nonsafety-related SSCs were included to a point beyond the nonsafety-related/safety-related interface to a base-mounted component, flexible connection, or to the end of the piping run in accordance with the guidance of NEI 95-10. NEI 95-10, Appendix F describes the use of "bounding criteria" as a method of determining the portion of nonsafety-related SSCs to be included within the scope of license renewal.

The staff was unable to determine whether equivalent anchors (such as a combination of supports in the three orthogonal directions) had been used, in addition to the bounding criteria (base-mounted component, flexible connection, or the end of the piping run) discussed in the LRA and the 10 CFR 54.4(a)(2) project report which described the AMR of nonsafety-related systems and components affecting safety-related systems. In RAI 2.1-2, dated July 10, 2006, the staff requested that the applicant provide information related to the method used to develop the structural boundary and whether equivalent anchors had been used in addition to the bounding criteria discussed in the LRA.

In its response, by letters dated August 10, 2006, October 17, 2006, and July 3, 2007 the applicant further described the process used to determine the structural boundaries for those nonsafety-related systems which provided limited structural support to safety-related systems. As part of the applicant's evaluation, isometric drawings of plant piping systems were reviewed where applicable to determine the location of structural boundaries. These isometric drawings were developed as part of the plant design process utilizing the results of piping stress analyses. No new analyses or isometric drawings were developed to support the license renewal process. Rather, the existing drawings and analyses were used to develop the structural boundaries, and in those instances where isometric drawings were not readily available, the applicant used the bounding criteria in NEI 95-10 to identify the portions of the nonsafety-related system necessary to support the intended function. With respect to the use of equivalent anchors, the applicant stated that other than the actual structural boundaries identified as a result of the existing piping stress analysis, isometric drawings, and use of the bounding criteria, they did not use any equivalent anchors to identify the structural boundaries for the nonsafety-related systems identified as performing a 10 CFR 54.4(a)(2) function.

The staff reviewed the additional information provided by the applicant and found that the applicant has adequately addressed the staff's RAI based on the detailed description of the process used to identify the structural boundaries and confirmation that equivalent anchors were not used for the purposes of identifying structural boundaries for the nonsafety-related systems identified as performing a 10 CFR 54.4(a)(2) function. Therefore, the staff's concern described in RAI 2.1-2 is resolved.

Nonsafety-Related SSCs with a Potential for Spatial Interation with Safety-Related SSCs

The applicant considered physical impact, and fluid leakage, spray or flooding when evaluating the potential for spatial interaction between nonsafety-related systems and safety-related SSCs. The applicant used a spaces approach for scoping of nonsafety-related systems with potential spatial interaction with safety-related SSCs. The spaces approach focused on the interaction between nonsafety-related and safety-related SSCs that are located in the same space. A "space" was defined as a room or cubicle that is separated from other spaces by substantial objects (such as wall, floors, and ceilings). The space was defined such that any potential interaction between nonsafety-related and safety-related SSCs is limited to the space.

The 10 CFR 54.4(a)(2) project report stated that the applicant had evaluated situations where missiles could be generated from internal or external events such as failure of rotating equipment. The nonsafety-related design features that protect safety-related SSCs from such missiles are within the scope of license renewal. In addition, the 10 CFR 54.4(a)(2) project report stated that the applicant had evaluated overhead-handling systems to identify those whose structural failure could result in damage to any system that could prevent the accomplishment of a safety function. Nonsafety-related overhead-handling equipment determined to have a possible impact on safety-related SSCs were included within the scope of license renewal.

The LRA stated that the applicant had evaluated nonsafety-related portions of high-energy lines, including review of the UFSAR and relevant topical design basis document. As discussed in the 10 CFR 54.4(a)(2) project report, the applicant used these references to evaluate the high-energy lines for postulated pipe breaks and identified eleven systems within the reactor building and five systems outside the reactor building. The applicant's high-energy systems were evaluated to ensure identification of components that are part of nonsafety-related high-energy lines that can effect safety-related equipment. If the applicant's high-energy line break (HELB) analysis assumed that a nonsafety-related piping system did not fail, or assumed failure only at specific locations, then that piping system (piping, equipment and supports) was included in the scope of license renewal. Many of the identified systems were safety-related and included within the scope of license renewal in accordance with 10 CFR 54.4(a)(1). The remaining nonsafety-related high-energy lines that were determined to have potential interaction with safety-related SSCs were included within the scope of license renewal in accordance with 10 CFR 54.4(a)(2).

The applicant evaluated moderate and low energy systems that have the potential for spatial interactions of spray and leakage. Nonsafety-related systems and nonsafety-related portions of safety-related systems with the potential for spray or leakage that could prevent safety-related SSCs from performing their required safety function were considered in the scope of license renewal. In addition, the applicant evaluated retired in place (RIP) systems for potential for spatial interaction. These RIP systems include both air-filled and fluid-filled portions of systems which were depressurized and isolated or capped from the remaining system. The applicant performed a review of the material/environment combinations for the RIP systems to determine if leakage of any fluid-filled portions due to corrosion could create the potential for a spatial interaction. The applicant applied the guidance from the Electric Power Research Institute (EPRI), "Non-Class 1 Mechanical Implementation guideline and Mechanical Tools," Revision 4, 2006. Consistent with the EPRI tools guidance, the applicant determined that the current configuration of these systems would not provide the necessary mechanisms to cause a failure in these systems which could result in system degradation and the potential subsequent leakage.

The 10 CFR 54.4(a)(2) project report stated that the applicant used a "spaces" approach to identify the nonsafety-related SSCs which were located within the same space as safety-related SSCs. A space was defined as a room or cubicle, separated by walls, floors, and ceilings. The applicant documented the review of each mechanical system for potential spatial interaction with safety-related SSCs in applicant's scoping results report, as documented in the audit report. Following identification of the applicable mechanical systems, the applicant reviewed the system functions to determine whether the system contained fluid, air or gas. Nonsafety-related SSCs containing air or gas were excluded from the scope of license renewal. The applicant then reviewed the mechanical systems to determine whether the system had any components located within a safety-related structure. Those liquid-filled systems determined to have components located within a safety-related structure where then reviewed to determine if the system had components located within a space containing safety-related SSCs. Those nonsafety-related SSCs determined to contain fluid and to be located within a space containing safety-related SSCs were included within the scope license renewal.

In its letter dated July 3, 2007, the applicant included addition information in response to RAI 2.1-2 (which discussed nonsafety-related piping attached to safety-related SSCs). As a result of the staff's inspection activities, the applicant expanded its review of nonsafety-related SSCs located in the turbine building and the potential for spatial interaction with safety-related SSCs. The applicant identified that portions of certain systems within the scope of license renewal had been expanded to include additional nonsafety-related components located in the turbine building. These components are within the scope of license renewal due to the potential for spatial interaction with safety-related SSCs and are subject to an aging management review.

In addition, protective features, such as whip restraints, spray shields, supports, missile or flood barriers (which can prevent physical impact and fluid leakage, spray, or flooding), installed to protect safety-related SSCs against spatial interaction with nonsafety-related SSCs were included within the scope of license renewal.

2.1.4.2.3 Conclusion

Based on its review, the staff determines that the applicant's methodology for identifying systems and structures meets 10 CFR 54.4(a)(2) scoping criteria and, therefore, is acceptable. This determination is based on a review of sample systems, discussions with the applicant, and review of the applicant's scoping process.

2.1.4.3 Application of the Scoping Criteria in 10 CFR 54.4(a)(3)

2.1.4.3.1 Summary of Technical Information in the Application

In LRA Section 2.1.1.3, "Application of Criterion for Regulated Events," the applicant described the methodology for identifying systems, structures, and components relied on in safety analyses or plant evaluation to perform a function. Mechanical systems and structures that perform a intended function that demonstrates compliance with the regulations for fire protection (10 CFR 50.48), environmental qualification (10 CFR 50.49), pressurized thermal shock (10 CFR 50.61), anticipated transients without scram (ATWS) (10 CFR 50.62), and station blackout (SBO) (10 CFR 50.63) were included in the scope of license renewal. Mechanical systems and structures that have an intended function for 10 CFR 54.4(a)(3) are identified in LRA Sections 2.3 and 2.4. For example, LRA Section 2.3.2.2 states that the core spray (CS) system has two intended functions for 10 CFR 54.4(a)(3): the Appendix R safe shutdown capability analysis and the SBO coping analysis. LRA Section 2.4.3 states that the intake structure has one intended function for 10 CFR 54.4(a)(3): the Appendix R safe shutdown capability analysis for fire protection. All plant electrical and instrumental and control (EIC) systems and electrical equipment in mechanical systems were included in-scope of license renewal.

Fire Protection. The applicant described the scoping of mechanical systems and structures required to demonstrate compliance with the fire protection requirements in LRA Section 2.1.1.3.1, "Commission's Regulations for Fire Protection (10 CFR 50.48)." The applicant reviewed its CLB and identified the mechanical systems and structures relied upon to meet Appendix R and 10 CFR 50.48 requirements. Mechanical systems and structures credited with fire prevention, detection, mitigation in areas containing equipment important to safe operation of the plant, and equipment credited with safe shutdown in the event of a fire were included in-scope license renewal.

Environmental Qualification. The applicant described the environmental qualification requirements of 10 CFR 50.49 in LRA Section 2.1.1.3.2, "Commission's Regulations for Environmental Qualification (10 CFR 50.49)." All plant EIC systems and electrical equipment in mechanical systems were included in-scope of license renewal.

Pressurized Thermal Shock. These requirements are not applicable because Vermont Yankee is a Boiling Water Reactor.

Anticipated Transient Without Scram. The applicant described the scoping of mechanical systems and structures required to demonstrate compliance with the anticipated transient without scram (ATWS) requirements of 10 CFR 50.62 in LRA Section 2.1.1.3.4, "Commission's Regulations for Anticipated Transients without Scram (10 CFR 50.62)." Mechanical systems and structures that perform a 10 CFR 50.62 intended function were included in-scope of license renewal.

Station Blackout. The applicant described the scoping criteria in LRA Section 2.1.1.3.5, "Commission's Regulations for Station Blackout (10 CFR 50.63)." The applicants licensing basis requires a SBO coping duration of two hours and mechanical systems and structures required to support the two-hour coping duration are within the scope of license renewal. Although the switchyard is not considered a plant system, the offsite power system and related structures required to restore offsite power were also included in-scope of license renewal.

2.1.4.3.2 Staff Evaluation

The staff reviewed the applicant's approach to identifying mechanical systems and structures relied upon to perform a function related to the four regulated events applicable to boiling water reactors (BWRs) required by 10 CFR 54.4(a)(3). As part of this review, the staff discussed the methodology with the applicant, reviewed the documentation developed to support the review, and evaluated a sample of the resultant mechanical systems and structures identified as in-scope for 10 CFR 54.4(a)(3) criteria.

The LRPGs described the applicant's process for identifying systems and structures that are in the scope of license renewal. The LRPGs stated that all mechanical systems and structures that perform an intended function for 10 CFR 54.4(a)(3) are to be included in-scope of license renewal, and that the results of scoping are documented in the applicants scoping results report. The report also described the procedures and data base that were used to identify mechanical systems and structures for regulated events. In addition, the applicant used a variety of The Topical Design Basis Documents, as described in the audit report, to identify the principle systems for each regulated event. The applicants component database uses a classification code of "OQA" for components that are not safety-related but are subject to the requirements imposed by NRC regulations. Systems initially identified as not meeting the criterion of 10 CFR 54.4(a)(3) based on review of design basis information were reviewed for OQA components in the component database to verify that the systems performed no intended functions for license renewal regulated events.

Fire Protection. The applicant's LRPDs state that the Fire Hazard Analysis, Fire Protection and Appendix R Program, and Safe Shutdown Capability Analysis, are used to identify mechanical systems and structures that are in-scope of license renewal. The report identifies the mechanical systems that are included in-scope of license renewal because they perform a 10 CFR 50.48 intended function. For example, the fire protection system has one intended function, which is to extinguish fires in the vital areas of the plant. The LRPDs summarizes the scoping results for mechanical systems and identifies 23 mechanical systems which have one or more 10 CFR 50.48 intended functions. The report also identifies the structures that are included in-scope of license renewal because they perform a 10 CFR 50.48 function, and provides a summary of the scoping results for ten structures that have one or more 10 CFR 50.48 intended functions. For example, the carbon dioxide (CO_2) tank foundation has one intended function, which is to provide support for the CO_2 tank.

Environmental Qualification. For the environmental qualification regulated event, the staff reviewed the LRA, the applicant's implementation procedures, results reports, and the master equipment list. These were used by the applicant to identify environmental qualification components within the scope of license renewal. The staff also reviewed the environmental qualification list which was used by the applicant during the screening process to identify short-lived components.

Anticipated Transient Without Scram. The applicant's scoping results report identifies the mechanical systems that are included in-scope of license renewal because they perform a 10 CFR 50.62 intended function. For example, one intended function of the control rod drive (CRD) system is to provide alternate rod insertion during an ATWS event. The report summarizes the scoping results for mechanical systems, identifies that the CRD and SLC systems perform 10 CFR 50.62 intended functions, and identifies one structure that is included in-scope of license renewal because it performs a 10 CFR 50.62 intended function. A criterion for including the reactor building in-scope of licensee renewal was that it housed equipment credited for ATWS.

Station Blackout. The applicant's scoping results report states that mechanical systems and structures credited with the two-hour coping duration and switchyard components required to restore offsite power are included in-scope of license renewal. The report identifies the mechanical systems that are were included in-scope of license renewal because they perform a 10 CFR 50.63 intended function. For example, the CS system has one intended function which is to provide reactor coolant makeup in the SBO coping analysis. The report summarizes the scoping results for mechanical systems, identifies eight mechanical systems that have one or more 10 CFR 50.63 intended functions, and identifies that the Offsite Power system is in-scope of license renewal because it performs a 10 CFR 50.63 intended function. The report also identifies the structures that were included in-scope of license renewal because they perform a 10 CFR 50.63 function. For example, the Vernon Hydroelectric Station (VHS) had one intended function which is to maintain integrity for SBO. The report summarizes the scoping results for structures and identifies five structures that have one or more 10 CFR 50.63 intended functions.

Section 54.4(a)(3) of 10 CFR requires that all systems and structures relied on in safety analyses or plant evaluations to perform a function that demonstrates compliance with the Commission's regulation for SBO (10 CFR 50.63) be included in the scope of license renewal. LRA Section 2.1.1.3.5 stated that the VHS is credited as the alternate alternating current (AC) power source for SBO. LRA Section 2.4.5 states that the VHS structures are in-scope of license renewal. LRA Section 2.3.5 and the applicant's scoping results report identify the VHS structures that are in the scope of license renewal. However, the VHS mechanical and electrical systems were not explicitly identified as being included in the scope of license renewal. It was not clear to the staff why the Vernon Station mechanical and electrical systems were not identified in the scope of license renewal in accordance with 10 CFR 54.4(a)(3). Therefore, the staff submitted RAI 2.1-3 requesting that the applicant describe the scoping and screening methodology as it applies to the mechanical and electrical systems associated with the VHS, and identify those mechanical and electrical systems and components (SCs) that are in the scope of license renewal and subject to an AMR.

In its responses, by letters dated July 14, 2006, August 10, 2006, and October 20, 2006, the applicant further described the scoping and screening process used to evaluate the VHS. The applicant identified the VHS as the alternate alternating current source credited for the VYNPS loss of all alternating current power compliance with 10 CFR 50.63 (SBO rule), and therefore, in-scope of license renewal. The applicant stated, in part, that they had credited the Federal Energy Regulatory Commission dam inspection program to manage the effects of aging on the civil and structural elements of the VHS. All additional mechanical and electrical systems associated with the turbine generator (TG) were considered an active assembly that is routinely confirmed through normal operation and therefore, consistent with the screening process, determined to not be subject to an AMR. Notwithstanding the screening of the mechanical and electrical systems as part of the active assembly, the applicant performed an IPA of the passive, long-lived electrical and mechanical components of the VHS. On the basis of this evaluation, the applicant identified specific structural, mechanical, and electrical SSCs that support one or more of the intended functions of the VHS, which is consistent with the screening methodology described in Section 2.1.5.

The staff reviewed the applicant's responses to the RAI and concluded that the applicant has adequately described its process for scoping and screening of the VHS, and has identified the VHS as in-scope. The applicant has also evaluated the SSCs associated with the VHS, consistent with the screening methodology described in Section 2.1.5. The staff found that the applicant has adequately addressed the staff's RAI. Therefore, the staff's concern described in RAI 2.1-3 is resolved.

2.1.4.3.3 Conclusion

On the basis of the sample review, discussions with the applicant, the applicants RAI response, and review of the applicant's scoping process, the NRC staff determines that the applicant's methodology for identifying systems and structures meets the scoping criteria of 10 CFR 54.4(a)(3), and is therefore acceptable.

2.1.4.4 Plant-Level Scoping of Systems and Structures

2.1.4.4.1 Summary of Technical Information in the Application

System and Structure Level Scoping. The applicant documented its methodology for performing the scoping of SSCs in accordance with 10 CFR 54.4(a) in its LRPGs and LRPDs. The applicant's approach to system and structure scoping provided in the site guidance was consistent with the methodology described in LRA Section 2.1. The LRPGs specify that the personnel performing license renewal scoping use CLB documents, describe the system or structure, and list the functions that the system or structure is required to accomplish. Sources of information regarding the CLB for systems included the UFSAR, DBDs, VYNPS component database, Maintenance Rule scoping reports, control drawings, and docketed correspondence. The applicant then compared identified system or structures function lists to the scoping criteria to determine whether the functions met the scoping criteria of 10 CFR 54.4(a). The applicant documented the results of the plant-level scoping process in accordance with the LRPGs. These results were provided in the systems and structures LRPDs. The information in the LRPDs includes a description of the structure or system, a listing of functions performed by the system or structure, information pertaining to system realignment (as applicable), identification of intended functions, the 10 CFR 54.4(a) scoping criteria met by the system or structure, references, and the basis for the classification of the system or structure intended functions. During the scoping methodology audit, the staff reviewed a sampling of LRPD reports and concluded that the applicant's scoping results in the LRPDs contained an appropriate level of detail to document the scoping process.

Conclusion

On the basis of a review of the LRA, the scoping and screening implementation procedures, and a sampling review of system and structure scoping results during the methodology audit, the staff concludes that the applicant's scoping methodology for systems and structures was adequate. In particular, the staff determines that the applicant's methodology reasonably identified systems and structures within the scope of license renewal and their associated intended functions.

Component Level Scoping. After the applicant identified the systems and structures within the scope of license renewal, a review of mechanical systems and structures was performed to determine the components in each in-scope system and structure. The structural and mechanical components that supported intended functions were considered within the scope of license renewal and screened to determine if an AMR was required. All electrical components within the mechanical and electrical systems were included in-scope as commodity groups (groups of like structures and components). The applicant considered three component classifications during this stage of the scoping methodology: mechanical, structural, and electrical. The VYNPS component database and controlled plant drawings provide a comprehensive listing of plant components. Component type and unique component identification numbers were used to identify each component identified as in-scope and subject to an AMR.

Commodity Groups Scoping. Initially all electrical components within the mechanical and electrical systems were included in the scope of license renewal as commodity groups. Since many electrical component types are considered active in accordance with the guidance in NEI 95-10 and the SRP-LR, they were screened out as not meeting the passive criteria and were subsequently not subject to an AMR. In LRA Section 2.1.2.3, the applicant described the commodity groups used to evaluate all in-scope electrical components subject to an AMR.

Structural components were grouped as structural commodity types. Commodity types were based on materials of construction. LRA Section 2.1.2.2.1 identified the various structural commodity groups including:

- steel
- threaded fasteners
- concrete
- fire barriers

- elastomers
- earthen structures
- flouropolymers and lubrite sliding surfaces

Insulation. LRA Section 2.4.6, "Bulk Commodities," stated that insulation may have the specific intended functions of (1) controlling the heat load during design basis accidents in areas with safety-related equipment, or (2) maintaining integrity such that falling insulation does not damage safety-related equipment (reflective metallic type reactor vessel insulation). As such, insulation is included in the scope of license renewal as a commodity group in those applications where it provides one or both of the above intended function.

Consumables. In LRA Section 2.1.2.4, "Consumables," the applicant discussed consumables. The guidance in Table 2.1-3 in NUREG-1800 was used to categorize and evaluate consumables. Consumables were divided into the following four categories for the purpose of license renewal: (a) packing, gaskets, component seals, and O-rings; (b) structural sealants; (c) oil, grease, and component filters; and (d) system filters, fire extinguishers, fire hoses, and air packs.

The consumables in both categories (a) and (b) are considered as subcomponents. Category (a) subcomponents are not relied upon to form a pressure-retaining function and, therefore, not subject to an AMR. Category (b) subcomponents are structural sealants for structures within the scope of license renewal that require an AMR. Category (c) consumables are periodically replaced in accordance with plant procedures and, therefore, not subject to an AMR. Category (d) consumables are subject to replacement based on National Fire Protection Association (NFPA) standards in accordance with plant procedures and, therefore, not subject to an AMR.

2.1.4.4.2 Staff Evaluation

The staff reviewed the applicant's methodology for performing the scoping of plant systems and components to ensure it was consistent with 10 CFR 54.4(a). The methodology used to determine the mechanical systems and components in-scope of license renewal was documented in LRPDs and plant level scoping results were identified in LRA Table 2.2-1. The scoping process defined the entire plant in terms of systems and structures. As specified in the LRPGs, the applicant identified the systems and structures that are subject to 10 CFR 54.4 review, described the processes for capturing the results of the review, and determined if the system or structure performed intended functions consistent with the criteria of 10 CFR 54.4(a). The process was completed for all systems and structures to ensure that the entire plant was addressed. The applicant's technical personnel performed initial reviews on systems and structures identified in the CLB.

The staff noted that a system or structure was presumed to be in-scope of license renewal if it performed one or more safety-related functions or met the other scoping criteria per the Rule as determined by CLB review. Mechanical and structural component types that supported intended functions were considered in-scope of license renewal. All component types in electrical systems in-scope of license renewal were considered in-scope of license renewal. These component types were placed in commodity groups. The electrical commodity groups were further screened to determine if they required an AMR. The staff did not identify any discrepancies with the methodology used by the applicant.

The staff reviewed the methodology used by the applicant to generate commodity groups. Separate commodity groups were identified for various mechanical, structural, and electrical components and were identified in the LRPDs. The staff reviewed the commodity group level functions that were identified and evaluated by the applicant in accordance with 10 CFR 54.4(a). This process determined whether the commodity group was considered in-scope of license renewal. The staff found the methodology used acceptable.

The staff reviewed the results of the scoping process documented in the LRPDs in accordance with the LRPGs. This documentation included the description of the system or structure and the 10 CFR 54.4(a) scoping criteria met by the system or structure. The staff also reviewed a sample of the applicant's scoping documentation and concluded that it contained an appropriate level of detail to document the scoping process.

The staff reviewed the applicant's evaluation of plant insulation as documented in the LRPD and the bulk commodities AMR. The applicant identified insulation as being in-scope and subject to an AMR based on it providing intended functions of insulating characteristics to reduce heat transfer, and structural or functional support to nonsafety-related SCs whose failure could prevent safety-related functions. Both mirror and non-mirror insulation were evaluated. The staff concludes that the applicant's methods and conclusions regarding insulation are acceptable.

The staff reviewed the scoping and screening of consumables and finds that the applicant followed the process described in NUREG-1800, and appropriately identified and categorized the various consumables in accordance with the guidance. Plant consumables were initially identified and evaluated to determine if any met the criteria requiring an AMR, such as structural sealants. Additionally, the applicant identified all pertinent industry guidelines which were used as the basis for replacement of the item, such as NFPA standards.

2.1.4.4.3 Conclusion

Based on its review of the LRA, scoping and screening implementation procedures, and a sampling of system scoping results during the audit, the staff concludes that the applicant's scoping methodology for plant SSCs, commodity groups, insulation, and consumables is acceptable. In particular, the staff determines that the applicant's methodology reasonably identifies systems, structures, component types, and commodity groups within the scope of license renewal and their intended functions.

2.1.4.5 Mechanical Component Scoping

2.1.4.5.1 Summary of Technical Information in the Application

In LRA Section 2.1, the applicant described the methodology for identifying mechanical system components that are in the scope of license renewal. For mechanical systems, the mechanical components that support the system intended functions are included in the scope of license renewal. For mechanical system scoping, a system was defined as the collection of components in the component database assigned to the system code. System intended functions were determined based on the functions performed by those components. Defining a system by the components in the database is generally consistent with the VYNPS maintenance rule scoping documents and safety classification procedure. Each mechanical system was evaluated against the criteria of 10 CFR 54.4 to determine which system components performed the intended functions consistent with the scoping criteria.

2.1.4.5.2 Staff Evaluation

The staff evaluated LRA Section 2.1 and the guidance in LRPDs, LRPGs, and aging management (AM) reports to complete the review of mechanical scoping process. The program guidelines and AM reports provided instructions for identifying and evaluating individual mechanical system components with respect to the scoping criteria. The CLB documents were utilized when determining whether a system or component is within the scope of 10 CFR 54.4(a). Examples of these sources included, but were not limited to, the UFSAR, Maintenance Rule database, separate ATWS, environmental qualification, fire protection and SBO documents, technical specifications, safety evaluation reports. Additional sources of mechanical component information included the VYNPS component database and individual system flow diagrams.

Mechanical system diagrams were evaluated to create license renewal boundaries for each system showing the in-scope components. Components that support a safety-related function or a regulated event were identified and further evaluated during the screening process to determine if the component should be subject to an AMR. Nonsafety-related components that are connected to safety-related components and provide structural support at the safety/nonsafety interface, or components whose failure could prevent satisfactory accomplishment of a safety-related function due to spatial interaction with safety-related SSCs are included in-scope and individually identified in the AMR associated with the 10 CFR 54.4(a)(2) evaluation, but were not specifically highlighted on the license renewal drawings. As part of the applicant's verification process, the list of mechanical components identified as in-scope were compared to the data in LRIS and the VYNPS component database to confirm the scope of components in the system.

The staff reviewed the implementation guidance and the CLB documents associated with mechanical system scoping, and found that the guidance and CLB source information noted above were acceptable to identify mechanical components and support structures in mechanical systems that are within the scope of license renewal. The staff conducted detailed discussions with the applicant's license renewal project management personnel and reviewed documentation pertinent to the scoping process. The staff assessed whether the applicant had appropriately applied the scoping methodology outlined in the LRA and implementation procedures and whether the scoping results were consistent with CLB requirements. The staff determined that the applicant's proceduralized methodology was consistent with the description provided in the LRA Section 2.1 and the guidance contained in SRP-LR, Section 2.1, and was adequately implemented.

Scoping Methodology for the Core Spray System

In LRA Section 2.3.2.2, "Core Spray," the applicant provided the scoping and screening methodology results for SSCs within the CS system. The CS system is a safety-related system and is credited with mitigating the effects of a loss of coolant events. The CS system accomplishes the following scoping criteria associated with the Rule:

The CS system has the following intended functions for 10 CFR 54.4(a)(1):

- Provide injection of water following loss of reactor coolant
- Support primary containment isolation
- Provide reactor coolant pressure boundary

The CS system has the following intended function for 10 CFR 54.4(a)(2):

- Maintain integrity of nonsafety-related components such that no physical interaction with safety-related components could prevent satisfactory accomplishment of a safety function

The CS system has the following intended function for 10 CFR 54.4(a)(3):

- The CS system is credited in the Appendix R safe shutdown capability analysis (10 CFR 50.48)
- The CS system is credited in the SBO coping analysis (10 CFR 50.63)

The CS license renewal scoping boundary includes those portions of nonsafety-related piping and equipment that extend beyond the safety-related/nonsafety-related interface. The scoping results indicated that the CS contains six system functions within the scope of license renewal.

As part of the audit, The staff reviewed the applicant's methodology for identifying CS mechanical component type meeting the scoping criteria as defined in the Rule. The staff also reviewed the scoping methodology implementation procedures and discussed the methodology and results with the applicant. The staff confirmed that the applicant had identified and used pertinent engineering and licensing information in order to determine the CS mechanical component type required to be in-scope of license renewal. As part of the review process, the staff evaluated each system intended function identified for the CS system, the basis for inclusion of the intended function, and the process used to identify each of the system components credited with performing the intended function. The staff confirmed that the applicant had identified and highlighted system piping and instrumentation diagrams (P&IDs) to develop the system boundaries in accordance with the procedural guidance. The applicant was knowledgeable about the process and conventions for establishing boundaries as defined in the license renewal implementation procedures. Additionally, the staff confirmed that the applicant had independently confirmed the results in accordance with the governing procedures. Specifically, other license renewal personnel knowledgeable about the system had independently reviewed the marked-up drawings to ensure accurate identification of system intended functions. The applicant performed additional cross-discipline verification and independent reviews of the resultant highlighted drawings before final approval of the scoping effort.

2.1.4.5.3 Conclusion

Based on its review of the LRA, scoping implementation procedures, and the system sample and discussions with the applicant, the staff concludes that the applicant's methodology for identifying mechanical systems for 10 CFR 54.4(a) scoping criteria is acceptable.

2.1.4.6 Structural Component Scoping

2.1.4.6.1 Summary of Technical Information in the Application

In LRA Section 2.1, the applicant described the methodology for identifying structures that are in the scope of license renewal. All plant structures and SBO-related non-plant structures were initially identified. Structure intended functions were identified using CLB documents such as the UFSAR, the Maintenance Rule document for buildings and structures, safety classification procedures, the fire hazards analysis, and the safe shutdown capability assessment. Structures that have an intended function for 10 CFR 54.4(a) were included in the scope of license renewal and listed in LRA Table 2.2-3. Structures that were not in-scope of license renewal are listed in LRA Table 2.2-4. LRA Section 2.4 describes the scoping results for the individual structures that are in-scope of license renewal. For example, LRA Section 2.4.1 describes the intake structure's purpose and seismic classification. The intake structure was in-scope of license renewal because it provides supports, shelter and protection for safety and nonsafety-related systems within the scope of license renewal.

2.1.4.6.2 Staff Evaluation

The staff reviewed the applicant's approach for identifying structures relied upon to perform the functions as required by 10 CFR 54.4(a). As part of this review, the staff discussed the methodology with the applicant, reviewed the documentation developed to support the review, and evaluated the scoping results for several structures that were identified in-scope of license renewal.

The LRPGs describe the applicant's process for identifying structures that are in the scope of license renewal and state that all structures that perform an intended function are to be included in-scope of license renewal and that the scoping results are to be documented in the scoping results report. The scoping results report lists all the structures that were evaluated and also describes the procedures that were used to identify structures. In additional, the plant UFSAR, Maintenance Rule Document, Fire Hazards Analysis, and Safe Shutdown Capability Analysis were used to identify structures. The applicant's component database uses a classification code of "BLD" for structures, and a search of this data base was used to identify structures.

The staff reviewed the applicants implementation procedures and scoping results reports. Structural scoping was performed in a manner to ensure that all plant buildings, yard structures, and SBO related non-plant structures were considered. The scoping results report identified the intended functions for each structure required for compliance with one or more criteria of 10 CFR 54.4(a). The structural component intended functions were identified based on the guidance provided in NEI 95-10 and NUREG-1800. For structures, the evaluation boundaries were determined by developing a complete description of each structure with respect to the intended functions performed by the structure. The results of the review were documented in the scoping results report (which contains a list of structures, evaluation results for each of the 10 CFR 54.4(a) criteria for each structure, a description of structural intended functions, and source reference information for the functions).

The staff conducted detailed discussions with the applicant's license renewal team and reviewed documentation pertinent to the scoping process. The staff assessed if the scoping methodology outlined in the LRA and procedures were appropriately implemented and if the scoping results were consistent with CLB requirements. The staff also reviewed structural scoping evaluation results for the intake structure and VHS to verify proper implementation of the scoping process. Based on these audit activities, the staff did not identify any discrepancies between the methodology documented and the implementation results.

2.1.4.6.3 Conclusion

Based on its review of the LRA, the applicant's detailed scoping implementation procedures, and a sampling of structural scoping results, the staff concludes that the applicant's methodology for identification of structural component types within the scope of license renewal meets 10 CFR 54.4(a) requirements and, therefore, is acceptable.

2.1.4.7 Electrical Component Scoping

2.1.4.7.1 Summary of Technical Information in the Application

LRA Section 2.1.1, "Scoping Methodology" describes the scoping process associated with electrical systems and components. For the purposes of system level scoping, plant EIC systems were included in the scope of license renewal. EIC components in mechanical systems were included in the evaluation of electrical systems. LRA Section 2.1.1 refers to LRA Section 2.5, "Scoping and Screening Results: Electrical and Instrumentation and Control Systems," which further states that the default inclusion of plant electrical and instrumentation and controls (EIC) systems in the scope of license renewal reflects the method used for the scoping of electrical systems, which is different from the methods used for mechanical systems and structures. The approach used for EIC components was to include components in the review unless they were specifically screened out. When used with the plant spaces approach, this method eliminated the need for unique identification of every component and its specific location. This gave assurance that components were not excluded from an AMR.

2.1.4.7.2 Staff Evaluation

The staff evaluated LRA Sections 2.1.1 and 2.5 and the applicants implementing procedures and aging management reports, as documented in the audit report, governing the electrical scoping methodology. The scoping phase for electrical components began with placing all electrical components from plant systems within the scope of license renewal. In addition, any electrical components from non-plant systems that met the criteria for inclusion of 10 CFR 54.4(a) (such as components credited for SBO) were also included within the scope of license renewal. The staff determined that the data sources used for scoping included the EMPAC data base, the station single line drawing, and the cable design procurement specifications. The applicant gathered and sorted the collection of all electrical components from the data sources and assembled the data into word processing file, called the "scoping" file. The staff reviewed selected portions of the data sources and the resulting assemblage of the data contained in the "scoping" file. The staff selected components for validation. The applicant demonstrated the component location in the data source and how the component was included in the "scoping" file through implementation of the LRPGs.

2.1.4.7.3 Conclusion

Based on its review of the LRA, the applicant's detailed scoping implementation procedures, and a sampling of electrical scoping results, the staff concludes that the applicant's methodology for identification of electrical components within the scope of license renewal meets 10 CFR 54.4(a) requirements and, therefore, is acceptable.

2.1.4.8 Conclusion for Scoping Methodology

Based on its review of the LRA and the scoping implementation procedures, the staff determines that the applicant's scoping methodology is consistent with SRP-LR guidance and has identified SSCs within the scope of license renewal as required by 10 CFR 54.4(a)(1), (a)(2), and (a)(3). Therefore, the staff concludes that the applicant's methodology meets 10 CFR 54.4(a) requirements.

2.1.5 Screening Methodology

2.1.5.1 General Screening Methodology

After identifying systems and structures within the scope of license renewal, the applicant implemented a process for identifying SCs subject to an AMR in accordance with 10 CFR 54.21.

2.1.5.1.1 Summary of Technical Information in the Application

In LRA Section 2.1.2, "Screening Methodology," the applicant discussed the method of identifying components from in-scope systems and structures that are subject to an AMR. The screening process consisted of the following steps:

> Identification of components that are long-lived or passive for each in-scope mechanical system, structure and electrical commodity group.

> Identification of the license renewal intended function(s) for all mechanical and structural component types and electrical commodity groups.

Active components were screened out and therefore, did not require an AMR. The screening process also identified short lived components and consumables. The short lived components are not subject to an AMR. Consumables are a special class of items that include packing, gaskets, component seals, O-rings, oil, grease, component filters, system filters, fire extinguishers, fire hoses, and air packs. Structural sealants for structures were the only consumables in-scope of license renewal that require an AMR.

2.1.5.1.2 Staff Evaluation

Pursuant to 10 CFR 54.21, the Commission requires that each LRA must contain an IPA that identifies SCs within the scope of license renewal that are subject to an AMR. The IPA must identify components that perform an intended function without moving parts or a change in configuration or properties (passive), as well as components that are not subject to periodic replacement based on a qualified life or specified time period (long-lived). The IPA includes a description and justification of the methodology used to determine the passive and long-lived SCs, and a demonstration that the effects of aging on those SCs will be adequately managed so that the intended function(s) will be maintained in accordance with all design conditions imposed by the plant-specific CLB for the period of extended operation.

The staff reviewed the methodology used by the applicant to determine if mechanical and structural component types, and electrical commodity groups in-scope of license renewal should be subject to an AMR. The applicant implemented a process for determining which SCs were subject to an AMR as required by 10 CFR 54.21(a)(1). In LRA Section 2.1.2, the applicant discussed these screening activities as they related to the component types and commodity groups within the scope of license renewal.

The screening process evaluated these in-scope component types and commodity groups to determine which ones were long-lived and passive and therefore, subject to an AMR. The staff reviewed LRA Sections 2.3, 2.4, and 2.5 that provided the results of the process used to identify component types and commodity groups subject to an AMR. The staff also reviewed the screening results reports for the CS system and intake structure.

The applicant provided the staff with a detailed discussion of the processes used for each discipline and provided administrative documentation that described the screening methodology. Specific methodology for mechanical, electrical, and structural is discussed below.

2.1.5.1.3 Conclusion

Based on its review of the LRA, the screening implementation procedures, and a sampling of screening results, the staff determines that the applicant's screening methodology is consistent with SRP-LR guidance and capable of identifying passive, long-lived components within the scope of license renewal and subject to an AMR. The staff determines that the applicant's process for identifying component types and commodity groups subject to an AMR meets 10 CFR 54.21 requirements and, therefore, is acceptable.

2.1.5.2 Mechanical Component Screening

2.1.5.2.1 Summary of Technical Information in the Application

In LRA Section 2.1.2.1, "Screening of Mechanical Systems," the applicant discussed the screening methodology for identifying passive and long-lived mechanical components and their support structures that are subject to an AMR. License renewal drawings were prepared to indicate portions of systems that support system intended functions within the scope of License renewal (with the exception of those systems in-scope for 10 CFR 54.4(a)(2) for physical interactions, as discussed below). In addition, the drawings identify components that are subject to an AMR. Boundary flags are used in conjunction with safety-to-nonsafety class breaks to identify the system intended function boundaries. Boundary flags are noted on the drawings as system intended function boundaries. All components within these boundary flags and class breaks support system intended functions within the scope of license renewal. Components subject to an AMR (i.e., passive, long-lived components that support system intended functions) were highlighted to indicate that the component was subject to an AMR.

2.1.5.2.2 Staff Evaluation

The staff evaluated the mechanical screening methodology in LRA 2.1.2.1, "Screening of Mechanical Systems," the LRPDs, LRPGs, and the AMR reports, as documented in the audit report. The mechanical system screening process began with the results from the scoping process. The applicant reviewed each mechanical system flow diagram to identify passive and long-lived components. To identify system components required to perform a system intended function, the applicant generated a listing of mechanical system components based on information derived from controlled system diagrams and the VYNPS component database. The LRPGs and LRPDs discuss in detail how to (1) determine system boundaries, (2) indicate components within a specific flow path which are required for performance of intended functions, and (3) determine and identify system and interdisciplinary interfaces (e.g., mechanical/structural, mechanical/electrical, structural/electrical). These components were entered into the LRIS database. The applicant also reviewed components in the VYNPS component database to confirm that all system components were considered. In cases where the mechanical system flow diagrams did not provide sufficient detail, such as large vendor supplied components (e.g., compressors, emergency diesel generators (EDGs)), the applicant reviewed associated component drawings or vendor manuals as necessary to identify individual components.

The staff reviewed the results of the boundary evaluation and discussed the process further with the applicant. The staff confirmed that mechanical system evaluation boundaries were established for each system within the scope of license renewal. These boundaries were determined by mapping the pressure boundary associated with system-level license renewal intended functions onto the controlled system drawings. Mechanical component types were loaded into a scoping and screening database and further review was performed to ensure all component types were identified. If a component type was not already in the LRIS, the component type was created for use in the license database. A preparer and an independent reviewer performed a comprehensive evaluation of the boundary drawings to ensure the completeness and accuracy of the review results. As part of the evaluation, the applicant also benchmarked passive and long-lived components for a system against previous LRAs containing similar systems.

As part of the audit, the staff reviewed the methodology used by the applicant to identify SSCs which met the screening criteria of the Rule. The staff confirmed that the applicant had implemented and followed the screening guidance in the SRP-LR and NEI 95-10. The staff confirmed the applicant had developed sufficiently detailed procedures for the screening of mechanical systems, had implemented those procedures, and had adequately documented the results in the associated AMR reports.

Additionally, the staff reviewed the screening activities associated with the CS system. The staff reviewed the system intended functions and associated source documents identified for the system, the CS flow diagrams, and the associated results documented in the AMR report. The staff did not identify any discrepancies with the evaluation, and determined that the applicant has adequately followed the process documented in the LRPDs and adequately documented the results in the AMR reports.

2.1.5.2.3 Conclusion

Based on its review of the LRA, the screening implementation procedures, and a sample review of CS screening results, the staff determines that the applicant's mechanical component screening methodology is consistent with SRP-LR guidance. The staff concludes that the applicant's methodology for identification of passive, long-lived mechanical components within the scope of License renewal and subject to an AMR meets 10 CFR 54.21(a)(1) requirements.

2.1.5.3 Structural Component Screening

2.1.5.3.1 Summary of Technical Information in the Application

The applicant described the methodology used for structural screening in LRA Sections 2.1.2.2, "Screening of Structures," and 2.4, "Scoping and Screening Results: Structures." LRA Section 2.1.2.2 states that structural components were evaluated to determine those subject to an AMR for each structure within the scope of license renewal. Specific structural components were identified from reviewing the CLB (drawings, etc.). Passive and long-lived structural components that performed an intended function were identified and subject to an AMR. NUREG-1800 and NEI 95-10, Appendix B, were used as the basis for the identification of passive structural components. Structural components (door, gate, pipe support, strut, or siding for example) were categorized as steel, threaded fasteners, concrete, fire barriers, elastomers, earthen structures, or flouropolymers and lubrite sliding surfaces. LRA Section 2.4 summarizes the screening results for structures. For example, LRA Section 2.4.3 and Table 2.4-3 summarize the screening results for the intake structure. LRA Section 2.4.5 and Table 2.4-5 summarize the screening results for the VHS. The structural components common to all structures such as piping supports were categorized as bulk commodities. LRA Section 2.4.6 and Table 2.4-6 summarize the screening results for structural bulk commodities.

2.1.5.3.2 Staff Evaluation

The staff reviewed the applicant's methodology for identifying structural components that are subject to an AMR as required in 10 CFR 54.21(a)(1). As part of this review, the staff discussed the methodology with the applicant, reviewed the documentation developed to support the activity, and evaluated the screening results for several structures that were identified in-scope of license renewal.

The applicant's aging management (AM) reports, as described in the audit report, provided detailed implementation guidance on the applicant's process for identifying and screening structural components that are subject to an AMR. The report stated that all structural components that perform an intended function and are passive and long-lived are subject to an AMR. In addition, the screening results for each system were described in separate AM reports for each system.

The staff reviewed the applicant's methodology used for structural screening described in LRA sections noted above, and in applicants implementing guidance and AM reports The applicant performed the screening review in accordance with the implementation guidance and captured pertinent structure design information, component, materials, environments, and effects of aging. The staff confirmed that the applicant used the lists of passive SCs embodied in the regulatory guidance as an initial starting point and supplemented that list with additional items unique to the site or for which a direct match to the generic lists did not exist (i.e., material/environment combinations). As one of the general rules for structural screening, the applicant determined that components which support or interface with electrical components such as, cable trays, conduits, instrument racks, panels and enclosures, were assessed as structural components.

The boundary for a structure was the entire building including base slabs, foundations, walls, beams, slabs, and steel superstructure. The AM reports identified each individual SC and indicated if the component is subject to an AMR. Each component was identified as a component, as a component type (door, gate, anchor support, strut, or siding for example), or as a material. The applicant provided the staff with a detailed discussion that described the screening methodology, as well as the screening reports for a selected group of structures.

The staff also examined the applicant's results from the implementation of this methodology by reviewing several of the plant structures (intake structure and VHS) identified as being in-scope. As part of this review, the staff reviewed the AM reports to verify that the applicant had performed a comprehensive evaluation and had identified the relevant structural components as part of their evaluation. The review included the evaluation of in-scope components, the corresponding component-level intended functions, and the resulting list of components subject to an AMR. The staff also discussed the process and results with the applicant. The staff did not identify any discrepancies between the methodology documented and the implementation results.

2.1.5.3.3 Conclusion

Based on its review of the LRA, the applicant's detailed screening implementation procedures, and a sampling of structural screening results, the staff concludes that the applicant's methodology for identification of passive, long-lived structural component types within the scope of License renewal and subject to an AMR meets 10 CFR 54.21(a)(1) requirements.

2.1.5.4 Electrical Component Screening

2.1.5.4.1 Summary of Technical Information in the Application

In the LRA Section 2.1.2.3, "Screening of Electrical and Instrumentation and Control Systems," the applicant discussed the use of NEI 95-10, Appendix B, "Typical Structure, Component and Commodity Groupings and Active/ Passive Determinations for the Integrated Plant Assessment," which identifies electrical commodities considered to be passive. The electrical commodity groups were identified and cross-referenced to the appropriate NEI 95-10 commodity.

The applicant determined that the majority of EIC commodity groups are active and do not require an AMR. Two passive EIC commodity groups were identified that meet the 10 CFR 54.21(a)(1)(i) criterion (components that perform an intended function without moving parts or without a change in configuration or properties):

- high-voltage insulators, and
- cables and connections, bus, electrical portions of EIC penetration assemblies

Additionally, the pressure boundary function that may be associated with some EIC components identified in NEI 95-10, Appendix B (flow elements, vibration probes) was considered in the mechanical AMRs, as applicable. Electrical components supported by structural commodities (cable trays, conduit and cable trenches) were included in the structural AMRs.

The applicant reviewed the passive electrical components to determine those components that were replaced based on a qualified life and therefore not subject to an AMR. The applicant determined that the components included in the Environmental Qualification of Electric Components Program per 10 CFR 50.49 are replaced based on qualified life and, therefore are not subject to an AMR. The applicant determined that the AMRs would be performed for the identified passive, non-Environmental Qualification EIC components.

2.1.5.4.2 Staff Evaluation

The staff reviewed the applicant's methodology used for electrical screening in LRA Sections 2.1.2.3 and the applicants implementation procedures and AM reports. The applicant used the screening process described in these documents to identify the electrical commodity groups subject to an AMR. The applicant used the VYNPS component database, the stations single line drawings, and cable procurement specifications as data sources to identify the EIC components including fuses-holders. The applicant determined there were no fuse-holders located outside of active devices and subject to an AMR.

The staff determined that the applicant had performed screening by initially identifying passive SCs and subsequently identifying the long-lived SCs contained within the passive SC population. The applicant identified seven commodities that were determined to meet the passive criteria. The seven commodities were further grouped in accordance with NEI 95-10 as (1) cables and connections, electrical portions of penetration assemblies, switchyard bus, transmission bus, transmission conductors and uninsulated ground conductors, and (2) high-voltage insulators. All were included in the "passive component table." The applicant then evaluated the passive commodities contained in the "passive component table" to identify whether they were subject to period replacement based on a qualified life or specified time period (short-lived), or not subject to period replacement based on a qualified life or specified time period (long-lived). The information used to identify short-lived components, which would not be subject to an AMR, included the environmental qualification master list. The environmental qualification master list identified the short-lived components included in the Environmental Qualification program. The remaining passive, long-lived components were included in the "passive, long-lived component table" and were determined to be subject to an AMR.

The staff reviewed the information contained in the scoping file, including the "passive component table," and the "passive, long-lived component table," to verify that the applicant had appropriately identified the identified those passive components which were long-lived and not subject to periodic replacement and therefore subject to an AMR. The staff reviewed the screening of selected components to verify the correct implementation of the LRPGs and AM reports.

2.1.5.4.3 Conclusion

The staff reviewed the LRA, procedures, electrical drawings, and a sample of the results of the screening methodology. The staff determines that the applicant's methodology was consistent with the description provided in LRA and the applicant's implementing procedures. On the basis of a review of information contained in the LRA, the applicant's screening implementation procedures, and a sampling review of electrical screening results, the staff concludes that the applicant's methodology for identification of electrical commodity groups subject to an AMR is consistent with the requirements of 10 CFR 54.21(a)(1), and is therefore acceptable.

2.1.5.5 Conclusion for Screening Methodology

Based on its review of the LRA, the screening implementation procedures, discussions with the applicant's staff, and a sample review of screening results, the staff determines that the applicant's screening methodology is consistent with the guidance of the SRP-LR and has identified passive, long-lived components within the scope of license renewal and subject to an AMR. The staff concludes that the applicant's methodology is consistent with the requirements of 10 CFR 54.21(a)(1) and, therefore, acceptable.

2.1.6 Summary of Evaluation Findings

The information in LRA Section 2.1, the supporting information in the scoping and screening implementation procedures and reports, and the information presented during the scoping and screening methodology audit and the applicant's responses to the staff's RAIs dated August 10, 2006, formed the basis of the staff's determination that the applicant's scoping and screening methodology was consistent with the requirements of the Rule. Based on this determination, the staff concludes that the applicant's methodology for identifying SSCs within the scope of license renewal and SCs requiring an AMR is consistent with the requirements of 10 CFR 54.4 and 10 CFR 54.21(a)(1), and, therefore, acceptable.

2.2 Plant-Level Scoping Results

2.2.1 Introduction

In LRA Section 2.1, the applicant described the methodology for identifying SSCs within the scope of License renewal. In LRA Section 2.2, the applicant used the scoping methodology to determine which SSCs must be included within the scope of License renewal. The staff reviewed the plant-level scoping results to determine whether the applicant has properly identified all systems and structures relied upon to mitigate DBEs, as required by 10 CFR 54.4(a)(1), systems and structures the failure of which could prevent satisfactory accomplishment of any safety-related functions, as required by 10 CFR 54.4(a)(2), and systems and structures relied on in safety analyses or plant evaluations to perform functions required by regulations referenced in 10 CFR 54.4(a)(3).

2.2.2 Summary of Technical Information in the Application

In LRA Tables 2.2-1a, 2.2-1b, and 2.2.3, the applicant listed plant mechanical systems, structures, and EIC systems, respectively, within the scope of license renewal. In LRA Tables 2.2-2 and 2.2-4, the applicant listed mechanical systems and structures that are not within the scope of license renewal. Based on the DBEs considered in the plant's CLB, other CLB information relating to nonsafety-related systems and structures, and certain regulated events, the applicant identified plant-level systems and structures within the scope of license renewal as specified by 10 CFR 54.4.

2.2.3 Staff Evaluation

In LRA Section 2.1, the applicant described its methodology for identifying systems and structures within the scope of license renewal and subject to an AMR. The staff reviewed the scoping and screening methodology and provides its evaluation in SER Section 2.1. To verify that the applicant properly implemented its methodology, the staff's review focused on the implementation results shown in LRA Tables 2.2-1a, 2.2-1b, 2.2-2, 2.2-3, and 2.2-4, to confirm that there were no omissions of plant-level systems and structures within the scope of license renewal.

The staff determined whether the applicant properly identified the systems and structures within the scope of license renewal in accordance with 10 CFR 54.4. The staff reviewed selected systems and structures that the applicant had not identified as falling within the scope of license renewal to verify whether the systems and structures have any intended functions requiring their inclusion within the scope of license renewal. The staff's review of the applicant's implementation was conducted in accordance with the guidance in SRP-LR Section 2.2, "Plant-Level Scoping Results."

In LRA Section 2.2, the staff identified areas in which additional information was necessary to complete the review of the applicant's plant-level scoping results. The applicant responded to the staff's RAIs as discussed below.

LRA Table 2.2-4, "Structures Not within the Scope of License Renewal," identifies the office building (administration and service buildings) as not within the scope of license renewal. The table identifies two UFSAR sections as references for office building. UFSAR Section 12.2.1.1.3 is an appropriate reference that identifies the administration building as a seismic Class II structure. However, the second UFSAR Section 12.2.3 is actually for the turbine building and not the administration or service building. In RAI 2.2-1 dated August 16, 2006, the staff requested that the applicant clarify and correct the reference to UFSAR Section 12.2.3 in LRA Table 2.2-4.

In its response dated September 20, 2006, the applicant stated that the office building is called by various names in VYNPS documents: the office building or area, the service building or area, and the administration building. It is sometimes considered part of the turbine building and in other contexts described as a separate building. In UFSAR Section 12.2.3, this area is listed as the "service area" that is part of the turbine building. Although the reference to UFSAR Section 12.2.3 is correct, this reference could have been omitted since UFSAR Section 12.2.3 only lists the service area and provides no description or further information about the service area. The applicant stated that the office building is not within the scope of license renewal.

Based on its review, the staff finds the applicant's response to RAI 2.2-1 acceptable because the applicant clarified the use of the term office building. Therefore, the staff's concern described in RAI 2.2-1 is resolved.

The pressure regulator and TG control system is described in USFAR Section 7.11. The purpose of the TG control system is to control steam flow and pressure to the turbine and to protect the turbine from overpressure or excessive speed. The TG controls work in conjunction with the "nuclear steam system" controls to maintain essentially constant reactor pressure and limit reactor transients during load variations. The LRA does not address the nuclear steam system, nor does it appear to refer to UFSAR Section 7.11 in the text. In RAI 2.2-3 dated August 16, 2006, the staff requested that the applicant clarify whether the nuclear steam system controls are included within the scope of license renewal, or explain the basis for their exclusion.

In its response dated September 20, 2006, the applicant stated that the pressure regulator and TG control system as described in UFSAR Section 7.11 is an electrical and instrumentation and control (EIC) portion of the main TG system listed in LRA Table 2.2-2. The TG system provides automatic and manual controls to maintain essentially constant reactor pressure and limit reactor transients during load variations. Components in the system control steam flow and pressure to protect the turbine from overpressure or excessive speed. As discussed in the introduction to Table 2.2-1b, "EIC Systems within the Scope of License Renewal (Bounding Approach)," all EIC commodities contained in electrical and mechanical systems are in-scope by default. LRA Table 2.2-1b provides the list of electrical systems that do not include mechanical components that meet the scoping criteria of 10 CFR 54.4. Systems (such as the TG system) with mechanical components that meet the scoping criteria of 10 CFR 54.4 are listed in LRA Table 2.2-1a. The pressure regulator and TG control system as described in UFSAR Section 7.11 are not considered separate systems and therefore are not listed in LRA Table 2.2-1a. However, the components that perform this function are in-scope as EIC components. The applicant stated that the nuclear steam system controls are within the scope of license renewal.

Based on its review, the staff finds the applicant's response to RAI 2.2-3 acceptable because the applicant stated all EIC commodities contained in electrical and mechanical systems are in-scope by default. Therefore, the staff's concern described in RAI 2.2-3 is resolved.

In response to concerns raised during the license renewal inspection, documented in the Vermont Yankee Nuclear Power Station - NRC License Renewal Inspection Report 05000271/2007006, dated June 4, 2007, the applicant placed fluid system components within the turbine building within the scope of license renewal. The applicants original scoping had determined that most of the turbine building was not within the scope of license renewal with a few exceptions, i.e., the diesel generator rooms, a few limited areas, and segments of the service water and diesel fuel oil systems. The inspection team determined that the scoping of segments of the service water and diesel fuel oil systems were not, in some instances, in accordance with guidance and that safety-related cables for reactor protection system functions had not been appropriately considered. The applicant added the turbine building to the scope of license renewal.

The applicant's response to the inspection report and subsequent submittal of supplementary information related to implementation of an enhanced scoping review are documented in the their letters to the NRC dated July 3, 2007, July 30, 2007, and August 16, 2007. As a result of implementing of scoping review changes, the applicant expanded the scope of license renewal and added the following mechanical systems and associated in-scope components:

- HD and HV instruments system
- air evacuation system
- building (drainage system components)system
- circulating water priming system
- extraction steam system
- heater drain system
- heater vent system
- hydrogen water chemistry system
- make-up demineralizer system
- seal oil system
- turbine building closed cooling water system
- main turbine generator
- turbine lube oil system

The above 13 mechanical systems were added to LRA Table 2.2-1a and removed from LRA Table 2.2-2.

The following mechanical systems had system boundary changes. For these systems, new component types were added that affected the scoping and screening results in the LRA. For systems listed below, new components, materials or environments that affected the AMR results in the LRA were added.

- augmented offgas system
- condensate system
- condensate demineralizer system
- condensate storage and transfer system

- circulating water system
- feedwater system
- fuel oil system
- fire protection system
- house heating boiler system
- heating, ventilation, and air conditioning system
- potable water system
- stator cooling system
- sampling system
- service water system

The effects of the above changes are evaluated in the applicable sections of this SER.

The staff reviewed the selected systems and structures that the applicant had not identified as falling within the scope of license renewal to verify whether the systems and structures have any intended functions that would require their inclusion within the scope of license renewal in accordance with 10 CFR 54.4. The staff's review of the applicant's implementation was conducted in accordance with the guidance described in SRP-LR Section 2.2, "Plant-Level Scoping Results."

2.2.4 Conclusion

The staff reviewed LRA Section 2.2, the RAI responses, the response to the license renewal inspection concerns, and the UFSAR supporting information to determine whether the applicant failed to identify any systems and structures within the scope of license renewal. The staff finds no such omissions. On the basis of its review, the staff concludes that there is reasonable assurance that the applicant has adequately identified in accordance with 10 CFR 54.4 the systems and structures within the scope of license renewal.

2.3 Scoping and Screening Results: Mechanical Systems

This section documents the staff's review of the applicant's scoping and screening results for mechanical systems. Specifically, this section discusses:

- reactor coolant system
- engineered safety features
- auxiliary systems
- steam and power conversion systems

In accordance with the requirements of 10 CFR 54.21(a)(1), the applicant's IPA must list passive, long-lived SCs within the scope of license renewal and subject to an AMR. To verify that the applicant properly implemented its methodology, the staff's review focused on the implementation results. This focus allowed the staff to confirm that there were no omissions of mechanical system components that meet the scoping criteria and are subject to an AMR.

The staff's evaluation of the information in the LRA was the same for all mechanical systems. The objective was to determine whether the applicant has identified, in accordance with 10 CFR 54.4, components and supporting structures for specific mechanical systems that appear to meet the license renewal scoping criteria. Similarly, the staff evaluated the applicant's screening results to verify that all passive, long-lived components were subject to an AMR in accordance with 10 CFR 54.21(a)(1).

In its scoping evaluation, the staff reviewed the applicable LRA sections and component drawings, focusing on components that have not been identified as within the scope of license renewal. The staff reviewed relevant licensing basis documents, including the UFSAR, for each mechanical system to determine whether the applicant has omitted from the scope of license renewal components with intended functions as required by 10 CFR 54.4(a). The staff also reviewed the licensing basis documents to determine whether the LRA specified all intended functions as required by 10 CFR 54.4(a). The staff requested additional information to resolve any omissions or discrepancies identified.

After its review of the scoping results, the staff evaluated the applicant's screening results. For those SCs with intended functions, the staff sought to determine whether: (1) the functions are performed with moving parts or a change in configuration or properties or (2) the SCs are subject to replacement after a qualified life or specified time period, as required by 10 CFR 54.21(a)(1). For those meeting neither of these criteria, the staff sought to confirm that these SCs were subject to an AMR, as required by 10 CFR 54.21(a)(1). The staff requested additional information to resolve any omissions or discrepancies identified.

Two-Tier Scoping Review Process for Balance of Plant (BOP) Systems

Of the 78 mechanical systems in the LRA, 44 are BOP systems which include most of the auxiliary systems and all of the steam and power conversion systems. The staff performed a two-tier scoping review for these BOP systems.

The two-tier scoping review process consists of Tier-1 and Tier-2 scoping reviews. The staff reviewed the LRA and UFSAR descriptions focusing on the system intended function to screen all the BOP systems into two groups based on the following screening criteria:

- safety importance/risk significance
- potential for system failure to cause failure of redundant safety system trains
- operating experience indicating likely passive failures
- systems subject to omissions based on previous LRA reviews

Examples of the safety important/risk significant systems are the instrument air (IA) system, the diesel generator (DG) and support systems, and the SW system, based on the results of the individual plant examination for VYNPS. An example of a system whose failure could result in common cause failure of redundant trains is a drain system providing flood protection. Examples of systems with operating experience indicating likely passive failures include MS system, feedwater system, and SW system. Examples of systems with identified omissions in previous LRA reviews include spent fuel cooling system and makeup water sources to safety systems.

From the 44 BOP systems, the staff selected 23 systems for a detailed "Tier-2" scoping review as described above. For the remaining 21 BOP systems, the staff performed a "Tier-1" scoping review of the LRA (which may have not included detailed boundary drawings) and UFSAR that would identify apparent missing components for an AMR. The following is a list of these 21 systems:

- service air (SA)
- SA and IA instruments
- condensate demineralizer
- RWCU filter demineralizer
- motor generator lube oil (MGLO)
- potable water
- equipment RIP
- stator cooling
- main steam, extraction steam and auxiliary steam instruments
- heater drain and heater vent (HD and HV) instruments
- air evacuation
- building (drainage system components)
- circulating water priming
- extraction steam
- heater drain
- heater vent
- make-up demineralizer
- seal oil
- turbine building closed cooling water
- main turbine generator
- turbine lube oil

The staff examined the applicant's environmental report in LRA Appendix E, Attachment E.1, "Evaluation of Probabilistic Safety Analysis Model," to verify that there is no risk significant system on the above list. None of the 21 systems is a significant contributor to the risk reduction worth rankings to core damage frequency or involved in the significant initiating events.

Systems Identified for Inspection

The staff used an inspection to verify 10 CFR 54.4(a)(2) scoping results. The staff identified several systems for the regional inspection team to include in its scoping and screening inspection. These systems had been included as within the scope of license renewal by the applicant as a result of the 10 CFR 54.4(a)(2) review. The staff requested that the inspection include a sampling review of the engineering report (if available), plant layout drawings and other documentation, and walkdowns of the plant areas that contain these systems and associated components. The systems identified for inspection include:

- augmented off-gas system
- circulating water system
- reactor water clean-up system

As a result of the regional inspection and other staff inquiry, the applicant issued letters to the NRC dated July 3, 2007, July 30, 2007, and August 16, 2007. These letters provided supplementary information that addressed resolution of the issues identified during the inspection. Refer to SER Sections 2.3.3.13A, 2.3.3.13E, and 2.3.3.13M for additional discussion.

2.3.1 Reactor Coolant System

LRA Section 2.3.1 states that the purposes of the reactor coolant system (RCS) are to house the reactor core and to contain and transport the fluids coming from or going to the reactor core. The RCS includes the reactor vessel and internals, the reactor recirculation system, CRD system, and Class 1 components that comprise the reactor coolant pressure boundary (RCPB), including MS and feedwater components. The applicant described the RCS as including the nuclear boiler (NB) system, the CRD system, and the hydraulic control unit (HCU) system associated with the CRDs.

The applicant described the supporting SCs of the RCS in the following LRA sections:

- 2.3.1.1   reactor vessel
- 2.3.1.2   reactor vessel internals
- 2.3.1.3   reactor coolant pressure boundary

The staff's findings on review of LRA Sections 2.3.1.1 - 2.3.1.3 are in SER Sections 2.3.1.1 - 2.3.1.3, respectively. The staff's review of the NB, CRD, and HCU systems proceeded as follows:

Summary of Technical Information in the Application. LRA Section 2.3.1 describes the RCS, including the NB, CRD, and HCU systems. Summaries of each system follow:

NB System. The NB system consists of Class 1 components, non-Class 1 components, and the following subsystems: reactor vessel and internals, reactor recirculation, MS, feedwater (Class 1); and nuclear boiler vessel instrumentation system (NBVIS). The reactor vessel is a welded vertical cylindrical pressure vessel with hemispherical heads. The cylindrical shell and hemispherical heads are fabricated of low-alloy steel plate. The vessel bottom head is welded directly to the vessel shell. The flanged upper head is secured to the vessel shell by studs and nuts. The reactor vessel includes nozzles, safe ends, CRD penetrations, instrument penetrations, and a support skirt. Additional details of the reactor vessel are described in LRA Section 2.3.1.1. The reactor vessel internals distribute the flow of coolant, locate and support the fuel assemblies, and provide an inner volume containing the core that can be flooded following a break in the nuclear system process barrier external to the reactor pressure vessel. Additional details of the reactor vessel internals are described in LRA Section 2.3.1.2.

Reactor recirculation provides a variable moderator (coolant) flow to the reactor core for adjusting reactor power level. Adjustment of the core coolant flow rate changes reactor power output, thus following plant load demand without adjusting control rods. The recirculation system is designed with sufficient fluid and pump inertia that fuel thermal limits cannot be exceeded as a result of recirculation system malfunctions. The reactor core is cooled by demineralized water which enters the lower portion of the core and boils as it flows upward around the fuel rods. The steam leaving the core is dried by steam separators and dryers in the upper portion of the reactor vessel, then directed to the turbine through four MS lines. The steam supply for high-pressure coolant injection (HPCI) and reactor core isolation cooling (RCIC) turbine operation is provided by connections to the MS piping. Class 1 feedwater lines provide water to the reactor vessel, entering near the top of the vessel downcomer annulus. Two feedwater lines divide and enter the vessel through four nozzles. Feedwater lines are also for injection of HPCI and RCIC. The NBVIS monitors reactor vessel parameters. The NBVIS is designed (1) to initiate and provide trip signals to interfacing plant safety systems, (2) to provide signals to interfacing plant nonsafety systems, and (3) to provide plant process parameter information necessary for normal, transient, and abnormal (including post-accident) operations. The NBVIS instrument sensing lines, including restriction orifices and excess flow check valves, are parts of the RCPB.

The NB system has safety-related components relied upon to remain functional during and following DBEs. The failure of nonsafety-related SSCs in the NB system could prevent the satisfactory accomplishment of a safety-related function. In addition, the NB system performs functions that support fire protection safe shutdown capability analysis and SBO coping analysis.

LRA Table 2.3.3-13-25 identifies the following nonsafety-related components types of the NB system within the scope of license renewal and subject to an AMR:

- bolting
- filter housing
- flow element
- orifice
- piping
- tubing
- valve body

The nonsafety-related NB system component intended function within the scope of license renewal is to provide a pressure boundary.

CRD System. The CRDs provide a means to control changes in core reactivity by incrementally positioning neutron-absorbing control rods within the reactor core in response to manual control signals. The CRD subsystem must shut down the reactor quickly (scram) by inserting control rods rapidly into the core in response to a manual or automatic signal.

The CRD system has safety-related components relied upon to remain functional during and following DBEs. The failure of nonsafety-related SSCs in the CRD system could prevent the satisfactory accomplishment of a safety-related function. In addition, the CRD system performs functions that support fire protection and ATWS.

LRA Table 2.3.3-13-5 identifies the following nonsafety-related CRD system component types within the scope of license renewal and subject to an AMR:

- bolting
- filter housing
- orifice
- piping
- pump casing
- strainer housing
- tank
- tubing
- valve body

The nonsafety-related CRD component intended function within the scope of license renewal is to provide a pressure boundary.

HCU System. The HCU system controls the water flow to the CRDs both for normal operation and during a reactor scram. Each HCU furnishes pressurized water upon signal to a CRD. The drive then positions its control rod as required. Water discharged from the drives during a scram flows through the HCUs to the scram discharge volume. Water discharged from a drive during a normal control rod positioning operation flows through its HCU and the exhaust header to the RWCU system discharge line.

The HCU system has safety-related components relied upon to remain functional during and following DBEs. The failure of nonsafety-related SSCs in the HCU system could prevent the satisfactory accomplishment of a safety-related function. In addition, the HCU system performs functions that support fire protection safe shutdown capability analysis and SBO coping analysis.

LRA Table 2.3.3-13-19 identifies the following nonsafety-related HCU system component types within the scope of license renewal and subject to an AMR:

- bolting
- filter housing
- piping
- tubing
- valve body

The nonsafety-related HCU system component intended function within the scope of license renewal is to provide a pressure boundary.

Staff Evaluation. The staff reviewed LRA Section 2.3.1, UFSAR Sections 3.4, 3.5, 4.1 through 4.6, and 7.18 using the evaluation methodology described in SER Section 2.3 and the guidance in SRP-LR Section 2.3, "Scoping and Screening Results: Mechanical Systems."

The staff evaluated the system functions described in the LRA and UFSAR to verify that the applicant had not omitted any components with intended functions from the scope of license renewal required by 10 CFR 54.4(a). The staff then reviewed those components that the applicant had identified as within the scope of license renewal to verify that no passive and long-lived components subject to an AMR had been omitted as required by 10 CFR 54.21(a)(1).

Conclusion. The staff reviewed the LRA to determine whether the applicant failed to identify any SSCs within the scope of license renewal or subject to an AMR. The staff finds no such omissions. On the basis of its review, the staff concludes that there is reasonable assurance that the applicant has adequately identified the NB, CRD, and HCU systems components within the scope of license renewal, as required by 10 CFR 54.4(a), and those subject to an AMR, as required by 10 CFR 54.21(a)(1).

2.3.1.1 Reactor Vessel

2.3.1.1.1 Summary of Technical Information in the Application

LRA Section 2.3.1.1 describes the reactor vessel, which contains the nuclear fuel core, core support structures, control rods, and other parts directly associated with the core. The major components of the reactor vessel are the reactor pressure vessel shell, bottom head, upper closure head, flanges, studs, nuts, nozzles and safe ends. The component evaluation boundaries are the welds between the safe ends and attached piping and the interface flanges for bolted connections. Thermal sleeves welded to vessel nozzles or safe ends, CRD stub tubes, CRD housings, incore housings, the vessel support skirt, and vessel interior and exterior welded attachments also were included.

LRA Table 2.3.1-1 identifies the following reactor vessel component types within the scope of license renewal and subject to an AMR:

- bolting
- heads and shell
- nozzles and penetrations
- safe ends, thermal sleeves, flanges, and caps
- vessel attachments and supports

The reactor vessel component intended functions within the scope of license renewal include the following:

- pressure boundary
- structural or functional support for safety-related equipment

2.3.1.1.2 Staff Evaluation

The staff reviewed LRA Section 2.3.1.1 and the UFSAR using the evaluation methodology described in SER Section 2.3 and the guidance in SRP-LR Section 2.3.

The staff evaluated the system functions described in the LRA and UFSAR to verify that the applicant has not omitted from the scope of license renewal any components with intended functions as required by 10 CFR 54.4(a). The staff then reviewed those components that the applicant has identified as within the scope of license renewal to verify that the applicant has not omitted any passive and long-lived components subject to an AMR as required by 10 CFR 54.21(a)(1).

In LRA Table 2.3.1-1, the reactor vessel leakage monitoring piping was not identified as a component within the scope of license renewal and requiring an AMR. In RAI 2.3.1.1-1 dated July 13, 2006, the staff requested that the applicant clarify whether the subject components were included within the scope of license renewal.

In its response dated August 15, 2006, the applicant stated that the subject components were included within the scope of license renewal in accordance with the category 'piping and fittings less than 4 inches NPS,' 'orifices (instrumentation),' and 'valve bodies less than 4 inches NPS' as part of RCPB components in Table 2.3.1-3. Based on its review, the staff finds the applicant's response to RAI 2.3.1.1-1 acceptable because the reactor vessel leakage monitoring piping was proven to be in-scope. The staff's concern described in RAI 2.3.1.1-1 is resolved.

In RAI 2.3.1.1-2 dated July 13, 2006, the staff requested that the applicant clarify if the scram discharge piping and volume are within the scope of license renewal because the subject components were not discussed in LRA Section 2.3.1.1.

In its response dated August 15, 2006, the applicant stated that the subject components were included within the scope of license renewal and subject to an AMR in accordance with the category 'piping and fittings less than 4 inches NPS,' 'orifices (instrumentation),' and 'valve bodies less than 4 inches NPS' as part of RCPB components in Table 2.3.1-3. Based on its review, the staff finds the applicant's response to RAI 2.3.1.1-2 acceptable because the scram discharge piping and volume were proven to be in-scope. The staff's concern described in RAI 2.3.1.1-2 is resolved.

In RAI 2.3.1.1-3 dated July 13, 2006, the staff requested that the applicant include the CRD housing supports within the scope of license renewal and requiring an AMR because the subject components were not discussed in LRA Section 2.3.1.1, "Reactor Vessel."

In its response dated August 15, 2006, the applicant stated that the subject components were considered in the category of structural elements and included in the line item for components and piping supports ASME Class 1, 2, 3 in Table 2.4-6, "Bulk Commodities Components Subject to an AMR." Based on its review, the staff finds the applicant's response to RAI 2.3.1.1-3 acceptable because the CRD housing supports were proven to be in-scope. The staff's concern described in RAI 2.3.1.1-3 is resolved.

2.3.1.1.3 Conclusion

The staff reviewed the LRA to determine whether the applicant failed to identify any SSCs within the scope of license renewal or subject to an AMR. The staff finds no such omissions. On the basis of its review, the staff concludes that there is reasonable assurance that the applicant has adequately identified the reactor vessel components within the scope of license renewal, as required by 10 CFR 54.4(a), and those subject to an AMR, as required by 10 CFR 54.21(a)(1).

2.3.1.2 Reactor Vessel Internals

2.3.1.2.1 Summary of Technical Information in the Application

LRA Section 2.3.1.2 describes the reactor vessel internals, which are designed to distribute the reactor coolant flow delivered to the vessel, to locate and support the fuel assemblies, and to contain the core in an inner volume that can be flooded following a break in the nuclear system process barrier. The reactor vessel internals are the control rod guide tubes, core plate, CS lines in the vessel, differential pressure and SLC line, feedwater spargers, fuel support pieces, incore guide tubes, incore dry tubes, local power range monitors, jet pump assemblies and jet pump instrumentation, shroud (including shroud stabilizers), shroud head and steam separator assembly, shroud support, steam dryer, surveillance sample holders, top guide, and vessel head spray line.

LRA Table 2.3.1-2 identifies the following reactor vessel internals component types within the scope of license renewal and subject to an AMR:

- control rod guide tubes
- core plate assembly
- core spray lines
- fuel support pieces
- incore dry tubes
- incore guide tubes
- jet pump assemblies
- jet pump casting
- shroud
- shroud repair hardware
- shroud support
- steam dryer
- top guide

The reactor vessel internals component intended functions within the scope of license renewal include the following:

- flow distribution

- boundary of a volume in which the core can be flooded and adequately cooled in the event of a breach in the nuclear system process barrier external to the reactor vessel

- pressure boundary

- structural or functional support for safety-related equipment

- structural integrity so loose parts are not introduced

2.3.1.2.2 Staff Evaluation

The staff reviewed LRA Section 2.3.1.2 and the UFSAR using the evaluation methodology described in SER Section 2.3 and the guidance in SRP-LR Section 2.3.

The staff evaluated the system functions described in the LRA and UFSAR to verify that the applicant has not omitted from the scope of license renewal any components with intended functions as required by 10 CFR 54.4(a). The staff then reviewed those components that the applicant has identified as within the scope of license renewal to verify that the applicant has not omitted any passive and long-lived components subject to an AMR as required by 10 CFR 54.21(a)(1).

2.3.1.2.3 Conclusion

The staff reviewed the LRA to determine whether the applicant failed to identify any SSCs within the scope of license renewal or subject to an AMR. The staff finds no such omissions. On the basis of its review, the staff concludes that there is reasonable assurance that the applicant has adequately identified the reactor vessel internals components within the scope of license renewal, as required by 10 CFR 54.4(a), and those subject to an AMR, as required by 10 CFR 54.21(a)(1).

2.3.1.3 Reactor Coolant Pressure Boundary

2.3.1.3.1 Summary of Technical Information in the Application

LRA Section 2.3.1.3 describes the RCPB, which maintains a high-integrity pressure boundary and fission product barrier inside the primary containment and to the first isolation outside the primary containment. Class 1 piping attached to the vessel nozzles or safe ends, including the welded joints, Class 1 pumps, and Class 1 boundary isolation valves, are included in this review. Connected Class 2 piping not part of another AMR, including vents, drains, leakoff, sample lines, and instrumentation lines up to the transmitters, is included as far as necessary to complete the RCS pressure boundary.

LRA Table 2.3.1-3 identifies the following RCPB component types within the scope of license renewal and subject to an AMR:

- bolting (flanges, valves, etc.)
- condensing chambers
- detector (CRD)
- drive (CRD)
- driver mount (RR)
- filter housing (CRD)
- flow elements (RR), (SLC)
- orifices (instrumentation)
- piping and fittings < 4 inches NPS
- piping and fittings $\geq$ 4 inches NPS
- pump casing and cover (RR)
- pump cover thermal barrier (RR)

- restrictors (MS)
- rupture disc (CRD)
- tank (CRD accumulator)
- thermowell
- valve bodies < 4 inches NPS
- valve bodies ≥ 4 inches NPS

The RCPB component intended functions within the scope of license renewal include the following:

- flow control
- pressure boundary

2.3.1.3.2 Staff Evaluation

The staff reviewed LRA Section 2.3.1.3 and the UFSAR using the evaluation methodology described in SER Section 2.3 and the guidance in SRP-LR Section 2.3.

The staff evaluated the system functions described in the LRA and UFSAR to verify that the applicant has not omitted from the scope of license renewal any components with intended functions as required by 10 CFR 54.4(a). The staff then reviewed those components that the applicant has identified as within the scope of license renewal to verify that the applicant has not omitted any passive and long-lived components subject to an AMR as required by 10 CFR 54.21(a)(1).

2.3.1.3.3 Conclusion

The staff reviewed the LRA to determine whether the applicant failed to identify any SSCs within the scope of license renewal or subject to an AMR. The staff finds no such omissions. On the basis of its review, the staff concludes that there is reasonable assurance that the applicant has adequately identified the RCPB components within the scope of license renewal, as required by 10 CFR 54.4(a), and those subject to an AMR, as required by 10 CFR 54.21(a)(1).

2.3.2 Engineered Safety Features

In LRA Section 2.3.2, the applicant identified the SCs of the engineered safety features that are subject to an AMR for license renewal.

The applicant described the supporting SCs of the engineered safety features in the following LRA sections:

- 2.3.2.1   residual heat removal
- 2.3.2.2   core spray
- 2.3.2.3   automatic depressurization
- 2.3.2.4   high pressure coolant injection
- 2.3.2.5   reactor core isolation cooling
- 2.3.2.6   standby gas treatment
- 2.3.2.7   primary containment penetrations

The staff's review findings regarding LRA Sections 2.3.2.1 - 2.3.2.7 are presented in SER Sections 2.3.2.1 - 2.3.2.7, respectively.

2.3.2.1 Residual Heat Removal

2.3.2.1.1 Summary of Technical Information in the Application

LRA Section 2.3.2.1 describes the RHR system, which removes decay heat energy from the reactor during both operational and accident conditions. The RHR system consists of two closed loops, each with two pumps in parallel, one heat exchanger, and the necessary valves and instrumentation. The RHR heat exchanger in each loop is cooled by the residual heat removal service water (RHRSW) system. The RHR system has eight modes of operation: (1) the low-pressure coolant injection (LPCI) mode takes suction from the suppression pool and injects flow into the core region of the reactor vessel through one of the two reactor recirculation loops to restore and maintain the water level of the reactor vessel following a loss of coolant accident (LOCA), (2) the containment spray cooling mode takes suction from the suppression pool and injects flow into spray headers located in the drywell and suppression chamber to reduce containment pressure and temperature following a LOCA by cooling any non-condensables and condensing any steam present, (3) the suppression pool cooling mode takes water from the suppression pool, passes it through the RHR heat exchangers, and returns flow to the suppression pool to remove heat added to the suppression pool, (4) the shutdown cooling mode takes water from the reactor vessel via the reactor recirculation A loop suction piping, passes it through the RHR heat exchangers, and returns flow to the reactor through the recirculation lines to remove sensible and decay heat from the reactor during shutdown, (5) the alternate shutdown cooling mode provides a cooling path if the normal shutdown cooling path is inoperable and can be initiated from the control room. RHR pumps take water from the suppression pool, pass it through RHR heat exchangers and inject into the vessel via RHR injection valves. Relief valves on the steam lines are open to allow overflow to the suppression pool, (6) the augmented fuel pool cooling (FPC) mode takes water from the FPC system, passes it through RHR heat exchangers, and returns flow to the FPC system to assist in FPC during reactor shutdown periods and the alternate cooling mode of operation and is not a safety function of RHR, (7) the emergency reactor vessel fill mode, which is beyond the design basis mode of operation, provides a cross-tie between the RHRSW system and RHR piping loop A. The RHRSW pumps take suction from the SW system and inject flow into the reactor vessel through RHR piping to provide a source of water to keep the reactor core covered (and fill containment) in the event that core standby cooling system (CSCS) pumps are lost due to loss of containment pressure or adequate core cooling cannot be assured, and (8) the alternate shutdown mode uses the RHR alternate shutdown panel to control the minimum valving required for vessel injection, torus cooling, and shutdown cooling modes to achieve and maintain cold shutdown conditions during a postulated control room or cable vault fire which eliminates normal means of system control.

The RHR system has safety-related components relied upon to remain functional during and following DBEs. The failure of nonsafety-related SSCs in the RHR system potentially could prevent the satisfactory accomplishment of a safety-related function. In addition, the RHR system performs functions that support fire protection safe shut down capability analysis.

LRA Tables 2.3.2-1 and 2.3.3-13-33 identify the following RHR system component types within the scope of license renewal and subject to an AMR:

- bolting
- cyclone separator
- heat exchanger (bonnet)
- heat exchanger (shell)
- heat exchanger (tubes)
- nozzle
- orifice
- piping
- pump casing
- strainer
- tank
- thermowell
- tubing
- valve body

The RHR system component intended functions within the scope of license renewal include the following:

- flow control
- filtration
- heat transfer
- pressure boundary

2.3.2.1.2 Staff Evaluation

The staff reviewed LRA Sections 2.3.2.1 and 2.3.3.13, and UFSAR Sections 4.8 and 6.4.4 using the evaluation methodology described in SER Section 2.3 and the guidance in SRP-LR Section 2.3.

The staff evaluated the system functions described in the LRA and UFSAR to verify that the applicant has not omitted from the scope of license renewal any components with intended functions as required by 10 CFR 54.4(a). The staff then reviewed those components that the applicant has identified as within the scope of license renewal to verify that the applicant has not omitted any passive and long-lived components subject to an AMR as required by 10 CFR 54.21(a)(1).

The LPCI coupling was identified in the Boiling Water Reactor Vessel and Internals Project (BWRVIP) -06 Report as a safety-related component. In RAI 2.3.2.1-1 dated July 13, 2006, the staff requested that the applicant identify LPCI couplings in the LRA as within the scope of license renewal and subject to an AMR if they are part of VYNPS.

In its response dated August 15, 2006, the applicant responded that VYNPS does not have LPCI couplings. Based on its review, the staff finds the applicant's response to RAI 2.3.2.1-1 acceptable because there are no LPCI couplings in-scope or subject to an AMR since there are no LPCI couplings at VYNPS. The staff's concern described in RAI 2.3.2.1-1 is resolved.

2-51

In RAI 2.3.2.1-2 dated July 13, 2006, the staff requested the applicant clarify whether vortex breakers are employed in the emergency core cooling system (ECCS) pump suction lines at VYNPS, and if so, identify and include these passive components in-scope requiring an AMR. In its response dated August 15, 2006, the applicant said that during the IPA for VYNPS, a review of site documentation for all in-scope mechanical systems, including licensing basis and DBDs, as well as the site component database and drawings was completed. The applicant determined that no vortex breakers were required to support system intended functions in the scope of license renewal per 54.4 (a)(1-3), and therefore, vortex breakers are not included in the LRA for VYNPS. Based on its review, the staff finds the applicant's response to RAI 2.3.2.1-2 acceptable because no vortex breakers support the intended function of the ECCS pump suction lines at VYNPS. The staff's concern described in RAI 2.3.2.1-2 is resolved.

2.3.2.1.3 Conclusion

The staff reviewed the LRA to determine whether the applicant failed to identify any SSCs within the scope of license renewal or subject to an AMR. The staff finds no such omissions. On the basis of its review, the staff concludes that there is reasonable assurance that the applicant has adequately identified the RHR system components within the scope of license renewal, as required by 10 CFR 54.4(a), and those subject to an AMR, as required by 10 CFR 54.21(a)(1).

2.3.2.2 Core Spray

2.3.2.2.1 Summary of Technical Information in the Application

LRA Section 2.3.2.2 describes the CS system, which in conjunction with other CSCS, provides adequate core cooling for all design basis break sizes up to and including double-ended breaks of the reactor recirculation system piping. The CS system protects the core in large breaks in the nuclear system when the RCIC and HPCI systems are unable to maintain reactor vessel water level. CS system protection also extends to small breaks in which the RCIC and HPCI systems are unable to maintain reactor vessel water level and automatic depressurization lowers reactor vessel pressure so the LPCI and the CS systems can cool the core. The CS system has two independent loops, each with a centrifugal water pump driven by an electric motor, a spray sparger in the reactor vessel above the core, and piping and valves to convey water from the suppression pool (primary safety-related source) or condensate storage tank (backup source) to the sparger.

The CS system has safety-related components relied upon to remain functional during and following DBEs. The failure of nonsafety-related SSCs in the CS system potentially could prevent the satisfactory accomplishment of a safety-related function. In addition, the CS system performs functions that support fire protection safe shutdown capability analysis and SBO coping analysis.

LRA Tables 2.3.2-2 and 2.3.3-13-6 identify the following CS system component types within the scope of license renewal and subject to an AMR:

- bolting
- bearing housing
- cyclone separator

- flow nozzle
- orifice
- piping
- pump casing
- strainer
- tubing
- valve body

The CS system component intended functions within the scope of license renewal include the following:

- flow control
- filtration
- pressure boundary

2.3.2.2.2 Staff Evaluation

The staff reviewed LRA Section 2.3.2.2 and UFSAR Sections 6.3 and 6.4.3 using the evaluation methodology described in SER Section 2.3 and the guidance in SRP-LR Section 2.3.

The staff evaluated the system functions described in the LRA and UFSAR to verify that the applicant has not omitted from the scope of license renewal any components with intended functions as required by 10 CFR 54.4(a). The staff then reviewed those components that the applicant has identified as within the scope of license renewal to verify that the applicant has not omitted any passive and long-lived components subject to an AMR as required by 10 CFR 54.21(a)(1).

2.3.2.2.3 Conclusion

The staff reviewed the LRA to determine whether the applicant failed to identify any SSCs within the scope of license renewal or subject to an AMR. The staff finds no such omissions. On the basis of its review, the staff concludes that there is reasonable assurance that the applicant has adequately identified the CS system components within the scope of license renewal, as required by 10 CFR 54.4(a), and those subject to an AMR, as required by 10 CFR 54.21(a)(1).

2.3.2.3 Automatic Depressurization

2.3.2.3.1 Summary of Technical Information in the Application

LRA Section 2.3.2.3 describes the automatic depressurization system (ADS), which actuates nuclear system pressure relief valves to depressurize the nuclear system automatically in a LOCA in which the HPCI system fails to deliver rated flow or break flow exceeds HPCI capacity (intermediate break). The depressurization of the nuclear system allows low-pressure standby cooling systems to supply enough cooling water to cool the fuel adequately. The ADS functions as one of the CSCSs. The ADS, in combination with the LPCI and CS systems, serves as a backup to the HPCI system.

The ADS has safety-related components relied upon to remain functional during and following DBEs. The failure of nonsafety-related ADS SSCs potentially could prevent the satisfactory accomplishment of a safety-related function. In addition, the ADS performs functions that support fire protection safe shutdown capability analysis and SBO coping analysis.

LRA Table 2.3.2-3 identifies the following ADS component types within the scope of license renewal and subject to an AMR:

- bolting
- orifice
- piping
- tubing
- valve body

The ADS component intended functions within the scope of license renewal include the following:

- flow control
- pressure boundary

2.3.2.3.2 Staff Evaluation

The staff reviewed LRA Section 2.3.2.3 and UFSAR Sections 4.4 and 6.4.2 using the evaluation methodology described in SER Section 2.3 and the guidance in SRP-LR Section 2.3.

The staff evaluated the system functions described in the LRA and UFSAR to verify that the applicant has not omitted from the scope of license renewal any components with intended functions as required by 10 CFR 54.4(a). The staff then reviewed those components that the applicant has identified as within the scope of license renewal to verify that the applicant has not omitted any passive and long-lived components subject to an AMR as required by 10 CFR 54.21(a)(1).

2.3.2.3.3 Conclusion

The staff reviewed the LRA to determine whether the applicant failed to identify any SSCs within the scope of license renewal or subject to an AMR. The staff finds no such omissions. On the basis of its review, the staff concludes that there is reasonable assurance that the applicant has adequately identified the ADS components within the scope of license renewal, as required by 10 CFR 54.4(a), and those subject to an AMR, as required by 10 CFR 54.21(a)(1).

2.3.2.4 High Pressure Coolant Injection

2.3.2.4.1 Summary of Technical Information in the Application

LRA Section 2.3.2.4 describes the HPCI system, which cools the reactor core adequately in a small break in the nuclear system with subsequent coolant loss which does not cause rapid depressurization of the reactor vessel. It performs this function simultaneously with a loss of normal auxiliary power. The HPCI system permits shutdown of the reactor by maintaining sufficient reactor vessel water inventory until the reactor vessel is depressurized. HPCI continues until reactor vessel pressure is below that at which the LPCI or CS system can maintain core cooling.

The HPCI system has safety-related components relied upon to remain functional during and following DBEs. The failure of nonsafety-related SSCs in the HPCI system potentially could prevent the satisfactory accomplishment of a safety-related function. In addition, the HPCI system performs functions that support fire protection and SBO coping analysis.

LRA Tables 2.3.2-4 and 2.3.3-13-20 identify the following HPCI system component types within the scope of license renewal and subject to an AMR:

- bearing housing
- bolting
- drain pot
- fan housing
- filter housing
- gear box
- governor housing
- heat exchanger (bonnet)
- heat exchanger (shell)
- heat exchanger (tubes)
- orifice
- piping
- pump casing
- sight glass
- steam trap
- strainer
- strainer housing
- tank
- thermowell
- tubing
- turbine casing
- valve body

The HPCI system component intended functions within the scope of license renewal include the following:

- flow control
- filtration
- heat transfer
- pressure boundary

2.3.2.4.2 Staff Evaluation

The staff reviewed LRA Sections 2.3.2.4 and 2.3.3.13, and UFSAR Sections 6.3 and 6.4 using the evaluation methodology described in SER Section 2.3 and the guidance in SRP-LR Section 2.3.

The staff evaluated the system functions described in the LRA and UFSAR to verify that the applicant has not omitted from the scope of license renewal any components with intended functions as required by 10 CFR 54.4(a). The staff then reviewed those components that the applicant has identified as within the scope of license renewal to verify that the applicant has not omitted any passive and long-lived components subject to an AMR as required by 10 CFR 54.21(a)(1).

2.3.2.4.3 Conclusion

The staff reviewed the LRA to determine whether the applicant failed to identify any SSCs within the scope of license renewal or subject to an AMR. The staff finds no such omissions. On the basis of its review, the staff concludes that there is reasonable assurance that the applicant has adequately identified the HPCI system components within the scope of license renewal, as required by 10 CFR 54.4(a), and those subject to an AMR, as required by 10 CFR 54.21(a)(1).

2.3.2.5 Reactor Core Isolation Cooling

2.3.2.5.1 Summary of Technical Information in the Application

LRA Section 2.3.2.5 describes the RCIC and the condensate storage and transfer (CST) systems. In the event of feedwater isolation with a simultaneous loss of normal auxiliary power, the RCIC system replaces the normal sources of makeup water to the reactor vessel to prevent uncovering of the core when it operates automatically without the use of any CSCSs. The RCIC system consists of a steam turbine-driven pump designed to supply water from either the condensate storage tank or the suppression pool to the reactor via the feedwater spargers. The purpose of the CST system is to provide a source of water to various plant systems, including the HPCI and RCIC systems (preferred source), CS system (as a backup source or for testing), the CRD system (backup source), and the spent fuel pool (fill and makeup source). The CST system connects to the condensate system to make up or draw off condensate to or from the hotwell. The CST system consists of the condensate storage tank, two condensate transfer pumps, piping, and valves.

The RCIC and CST systems have safety-related components relied upon to remain functional during and following DBEs. The failure of nonsafety-related SSCs in the system potentially could prevent the satisfactory accomplishment of a safety-related function. In addition, the systems perform functions that support fire protection safe shutdown capability analysis and SBO coping analysis.

LRA Tables 2.3.2-5, 2.3.3-13-7, and 2.3.3-13-31 identify the following RCIC and CST systems component types within the scope of license renewal and subject to an AMR:

- bolting
- condenser
- drain pot
- filter housing
- flow indicator
- heat exchanger (bonnet)
- heat exchanger (shell)

- heat exchanger (tubes)
- orifice
- piping
- pump casing
- rupture disk
- sight glass
- steam heater
- steam trap
- strainer
- strainer housing
- tank
- thermowell
- tubing
- turbine casing
- valve body

The component intended functions within the scope of license renewal include the following:

- flow control
- filtration
- heat transfer
- pressure boundary

2.3.2.5.2 Staff Evaluation

The staff reviewed LRA Sections 2.3.2.5 and 2.3.3.13, and UFSAR Sections 4.7 and 11.8.3.8 using the evaluation methodology described in SER Section 2.3 and the guidance in SRP-LR Section 2.3.

The staff evaluated the system functions described in the LRA and UFSAR to verify that the applicant has not omitted from the scope of license renewal any components with intended functions as required by 10 CFR 54.4(a). The staff then reviewed those components that the applicant has identified as within the scope of license renewal to verify that the applicant has not omitted any passive and long-lived components subject to an AMR as required by 10 CFR 54.21(a)(1).

2.3.2.5.3 Conclusion

The staff reviewed the LRA to determine whether the applicant failed to identify any SSCs within the scope of license renewal or subject to an AMR. The staff finds no such omissions. On the basis of its review, the staff concludes that there is reasonable assurance that the applicant has adequately identified the RCIC and CST systems components within the scope of license renewal, as required by 10 CFR 54.4(a), and those subject to an AMR, as required by 10 CFR 54.21(a)(1).

2.3.2.6 Standby Gas Treatment

2.3.2.6.1 Summary of Technical Information in the Application

LRA Section 2.3.2.6 describes the standby gas treatment (SBGT) system, which processes gaseous effluent from the primary and secondary containments when required to limit the discharge of radioactive materials to the environs and to limit ex-filtration from the secondary containment during primary containment isolation. This processing is accomplished by two trains, each capable of maintaining a negative pressure in the secondary containment and processing one net secondary containment volume of air per day through high-efficiency filters. The system functions as part of the secondary containment system. The SBGT system consists of two complete, independent trains, each a backup for the other and sized to handle the full system requirement. Each train has a demister, electric heaters, two high-efficiency particulate filters, a carbon absorber, a fan, and miscellaneous valves.

The SBGT system has safety-related components relied upon to remain functional during and following DBEs. The failure of nonsafety-related SSCs in the SBGT system potentially could prevent the satisfactory accomplishment of a safety-related function.

LRA Tables 2.3.2-6 and 2.3.3-13-38 identify the following SBGT system component types within the scope of license renewal and subject to an AMR:

- bolting
- duct
- fan housing
- filter
- filter housing
- filter unit housing
- orifice
- piping
- sight glass
- thermowell
- tubing
- valve body

The SBGT system component intended functions within the scope of license renewal include the following:

- filtration
- pressure boundary

2.3.2.6.2 Staff Evaluation

The staff reviewed LRA Section 2.3.2.6 and UFSAR Sections 1.6.2.15 and 5.3.4 using the evaluation methodology described in SER Section 2.3 and the guidance in SRP-LR Section 2.3.

The staff evaluated the system functions described in the LRA and UFSAR to verify that the applicant has not omitted from the scope of license renewal any components with intended functions as required by 10 CFR 54.4(a). The staff then reviewed those components that the applicant has identified as within the scope of license renewal to verify that the applicant has not omitted any passive and long-lived components subject to an AMR as required by 10 CFR 54.21(a)(1).

2.3.2.6.3 Conclusion

The staff reviewed the LRA to determine whether the applicant failed to identify any SSCs within the scope of license renewal or subject to an AMR. The staff finds no such omissions. On the basis of its review, the staff concludes that there is reasonable assurance that the applicant has adequately identified the SBGT system components within the scope of license renewal, as required by 10 CFR 54.4(a), and those subject to an AMR, as required by 10 CFR 54.21(a)(1).

2.3.2.7 Primary Containment Penetrations

2.3.2.7.1 Summary of Technical Information in the Application

LRA Section 2.3.2.7 describes the primary containment penetrations, which can rapidly isolate all pipes or ducts penetrating the primary containment with a containment barrier as effective as required to maintain leakage within permissible limits.

The primary containment penetrations have safety-related components relied upon to remain functional during and following DBEs.

LRA Table 2.3.2-7 identifies the following primary containment penetrations component types within the scope of license renewal and subject to an AMR:

- bolting
- piping
- valve body

The intended function of the primary containment penetrations is to provide a pressure boundary.

2.3.2.7.2 Staff Evaluation

The staff reviewed LRA Section 2.3.2.7 and UFSAR Sections 5.2.2, 5.2.3.4, and 5.2.3.5 using the evaluation methodology described in SER Section 2.3 and the guidance in SRP-LR Section 2.3.

The staff evaluated the system functions described in the LRA and UFSAR to verify that the applicant has not omitted from the scope of license renewal any components with intended functions as required by 10 CFR 54.4(a). The staff then reviewed those components that the applicant has identified as within the scope of license renewal to verify that the applicant has not omitted any passive and long-lived components subject to an AMR as required by 10 CFR 54.21(a)(1).

2.3.2.7.3 Conclusion

The staff reviewed the LRA to determine whether the applicant failed to identify any SSCs within the scope of license renewal or subject to an AMR. The staff finds no such omissions. On the basis of its review, the staff concludes that there is reasonable assurance that the applicant has adequately identified the primary containment penetrations components within the scope of license renewal, as required by 10 CFR 54.4(a), and those subject to an AMR, as required by 10 CFR 54.21(a)(1).

2.3.3 Auxiliary Systems

In LRA Section 2.3.3, the applicant identified the SCs of the auxiliary systems subject to an AMR for license renewal.

The applicant described the supporting SCs of the auxiliary systems in the following LRA sections:

- 2.3.3.1   standby liquid control
- 2.3.3.2   service water
- 2.3.3.3   reactor building closed cooling water
- 2.3.3.4   emergency diesel generator
- 2.3.3.5   fuel pool cooling
- 2.3.3.6   fuel oil
- 2.3.3.7   instrument air
- 2.3.3.8   fire protection-water
- 2.3.3.9   fire protection-carbon dioxide
- 2.3.3.10   heating, ventilation and air conditioning
- 2.3.3.11   primary containment atmosphere control/containment atmosphere dilution
- 2.3.3.12   John Deere diesel
- 2.3.3.13   miscellaneous systems in-scope for 10 CFR 54.4(a)(2)

The staff's review findings regarding LRA Sections 2.3.3.1 - 2.3.3.13 are presented in SER Sections 2.3.3.1 - 2.3.3.13, respectively.

2.3.3.1 Standby Liquid Control

2.3.3.1.1 Summary of Technical Information in the Application

LRA Section 2.3.3.1 describes the SLC system, which, independent of the control rods, shuts down the reactor from full power and maintains the reactor subcritical during cooldown. Maintaining subcriticality as the nuclear system cools assures that the fuel barrier is not threatened by overheating if not enough control rods can be inserted to counteract the positive reactivity effects of a colder moderator. The system, located in the reactor building, consists of a boron solution tank, a test water tank, two positive-displacement pumps, two explosive valves, an ion exchanger, a flush pump, piping, and valves. The liquid is pumped into the reactor vessel and discharged near the bottom of the core shroud to mix with the cooling water rising through the core.

The SLC system has safety-related components relied upon to remain functional during and following DBEs. The failure of nonsafety-related SSCs in the SLC system potentially could prevent the satisfactory accomplishment of a safety-related function. In addition, the SLC system performs functions that support ATWS.

LRA Tables 2.3.3-1 and 2.3.3-13-40 identify the following SLC system component types within the scope of license renewal and subject to an AMR:

- bolting
- gauge
- heater
- orifice
- piping
- pump casing
- sight glass
- strainer housing
- tank
- thermowell
- tubing
- valve body

The SLC system component intended function within the scope of license renewal is to provide a pressure boundary.

2.3.3.1.2 Staff Evaluation

The staff reviewed LRA Sections 2.3.3.1 and 2.3.3.13, and UFSAR Section 3.8 using the evaluation methodology described in SER Section 2.3 and the guidance in SRP-LR Section 2.3.

The staff evaluated the system functions described in the LRA and UFSAR to verify that the applicant has not omitted from the scope of license renewal any components with intended functions as required by 10 CFR 54.4(a). The staff then reviewed those components that the applicant has identified as within the scope of license renewal to verify that the applicant has not omitted any passive and long-lived components subject to an AMR as required by 10 CFR 54.21(a)(1).

2.3.3.1.3 Conclusion

The staff reviewed the LRA to determine whether the applicant failed to identify any SSCs within the scope of license renewal. The staff finds no such omissions. In addition, the staff's review determined whether the applicant failed to identify any components subject to an AMR.

The staff finds no such omissions. On the basis of its review, the staff concludes that there is reasonable assurance that the applicant has adequately identified the SLC system components within the scope of license renewal, as required by 10 CFR 54.4(a), and those subject to an AMR, as required by 10 CFR 54.21(a)(1).

2.3.3.2 Service Water

2.3.3.2.1 Summary of Technical Information in the Application

LRA Section 2.3.3.2 describes the SW system and the RHRSW system. The purpose of the SW system is to provide cooling water to various normal and emergency operating loads. The SW system consists of two parallel headers which supply cooling water to the following turbine and reactor auxiliary equipment: a reactor building closed cooling water (RBCCW) heat exchanger, RHR corner room ventilation coolers, a DG cooler, and an RHR heat exchanger (via the RHRSW pumps and piping). Each header is supplied by two pumps. The standby fuel pool cooling (SBFPC) system normally is supplied from the SW Train B header. The header and cross tie can be configured to be fed from the A header with B secured. Other turbine and reactor auxiliary equipment is supplied from a line tied into both headers. The purpose of the RHRSW system is to transfer heat from the RHR system during normal operation and accident conditions. The RHRSW system consists of four RHRSW pumps, two RHR heat exchangers and piping, valves, and instrumentation necessary to ensure system operation. The RHRSW pumps are supplied from the SW system. The cooling water then is pumped through the RHR heat exchangers and returned to the SW system.

The SW and RHRSW systems have safety-related components relied upon to remain functional during and following DBEs. The failure of nonsafety-related SSCs in the system potentially could prevent the satisfactory accomplishment of a safety-related function. In addition, the systems perform functions that support fire protection.

LRA Tables 2.3.3-2, 2.3.3-13-34, and 2.3.3-13-42 identify the following SW and RHRSW system component types within the scope of license renewal and subject to an AMR:

- bolting
- coil
- expansion joint
- fan housing
- heat exchanger (bonnet)
- heat exchanger (shell)
- heat exchanger (tubes)
- heat exchanger (tubesheets)
- indicator
- orifice
- piping
- pump casing
- strainer
- strainer housing
- suction barrel

- thermowell
- tubing
- valve body

The component intended functions within the scope of license renewal include the following:

- flow control
- filtration
- heat transfer
- pressure boundary
- structural or functional support for safety-related equipment

2.3.3.2.2 Staff Evaluation

The staff reviewed LRA Sections 2.3.3.2 and 2.3.3.13, and UFSAR Sections 10.6, 10.7, and 10.8 using the Tier-2 evaluation methodology described in SER Section 2.3 and the guidance in SRP-LR Section 2.3.

In conducting its review, the staff evaluated the system functions described in the LRA and UFSAR to verify that the applicant has not omitted from the scope of license renewal any components with intended functions as required by 10 CFR 54.4(a). The staff then reviewed those components that the applicant has identified as within the scope of license renewal to verify that the applicant has not omitted any passive and long-lived components subject to an AMR as required by 10 CFR 54.21(a)(1).

The staff's review of LRA Section 2.3.3.2 identified areas in which additional information was necessary to complete the review of the applicant's scoping and screening results. The applicant responded to the staff's RAIs as discussed below.

The staff noted that license renewal drawing LRA-G-191159-SH-01-0, at location H-12, depicts pipe section 2"-SW- 566C within the scope of license renewal. Upstream from where 2"-SW-566C enters the reactor building from the outside, there is no drawing continuation to depict the license renewal boundary. In RAI 2.3.3.2a-1 dated August 16, 2006, the staff requested that the applicant provide information for the continuation of 2"-SW-566C to the license renewal boundary and justify the boundary locations with respect to the applicable requirements of 10 CFR 54.4(a).

In its response dated September 20, 2006, the applicant stated that pipe section 2"-SW-566C contains vacuum breakers to prevent water-hammer in the nonsafety-related portion of the SW system. The portion of this piping outside of the reactor building wall ends at this point. There is no continuation of this portion of the piping.

Based on its review, the staff found the applicant response to RAI 2.3.3.2a-1 acceptable because the applicant confirmed this section of piping ends outside the reactor building wall and does not continue on another drawing. This is a section of piping open to atmosphere immediately outside of the reactor building to allow air flow to the vacuum breakers depicted on pipe section 2"-SW-566C. Therefore, the staff concern described in RAI 2.3.3.2a-1 is resolved.

The staff noted that license renewal drawing LRA-G-191159-SH-01-0, at location H-11, drawing note 16 indicates pipe section 4"-SW-567 and its supports on the reactor building alternate cooling supply piping (where the vacuum breakers tie in) are seismic Class II for structural integrity. This pipe section from valve 23D through valves RBAC-1A, 1B, 1C and 1D is not shown within the scope of license renewal. Failure of this pipe could have an adverse effect on the intended pressure boundary function for the service water piping. In RAI 2.3.3.2a-2 dated August 16, 2006, the staff requested that the applicant provide additional information about why this section of pipe and components are not shown within the scope of license renewal and justify the boundary locations with respect to the applicable requirements of 10 CFR 54.4(a).

In its response dated September 20, 2006, the applicant stated that this portion of piping is included for 10 CFR 54.4(a)(2) since it provides structural support for the safety-related portion of the system. As described in LRA Section 2.1.2.1.3, portions of systems included as required by 10 CFR 54.4(a)(2) are not shown on license renewal drawings. However, as discussed in LRA Table 2.3.3.1 3-8 for the SW system, the components outside the safety class pressure boundary, while relied upon to provide structural/seismic support for the pressure boundary are in-scope and subject to an AMR. This includes the portion of line 4"-SW-567 required to provide structural support for the vacuum breakers. In addition, this piping and associated valves are included as required by 10 CFR 54.4(a)(2) due to spatial interaction from spray or leakage since the line is in the reactor building.

Based on its review, the staff found the applicant response to RAI 2.3.3.2a-2 acceptable because the applicant acknowledged this section of piping 4" SW-567 from valve 23D to RBAC-1A, 1B, 1C, and 1D is within the scope of license renewal. As described in LRA Section 2.1.2.1.3, portions of systems included for 10 CFR 54.4(a)(2) are not shown on LRA drawings. Although the applicant did not identify this section of piping as being within the boundary of license renewal on the drawing, the applicant confirmed it is within the scope based on the potential for physical interaction with safety-related systems in accordance with 10 CFR 54.4(a)(2). Therefore, the staff concern described in RAI 2.3.3.2a-2 is resolved.

The staff noted license renewal drawing LRA-G-191159-SH-01-0, at location D-5, depicts the license renewal boundary on the downstream side of flow control valve (FCV)-104-17A. The pipe section from FCV-104-17A to the safety class boundary designation flag located at valve 171A and to the intake screens is not shown within the scope of license renewal. Similarly, the pipe section from FCV-104-17 B, C, D, and E to valves 17B, C, D and E and to the intake screens is also not shown within the scope of license renewal. Failure of these sections of pipe could have an adverse effect on the intended pressure boundary function for the service water piping. In RAI 2.3.3.2a-3 dated August 16, 2006, the staff requested that the applicant provide additional information about why these sections of piping and components are not shown within the scope of license renewal and justify the boundary locations with respect to the applicable requirements of 10 CFR 54.4(a).

In its response dated September 20, 2006, the applicant stated that the license drawings only show the portions of the system with intended functions that meet the requirements of 10 CFR 54.4(a)(1) or (a)(3). As described in LRA Section 2.1.2.1.3, portions of systems included as required by 10 CFR 54.4(a)(2) are not shown on license renewal drawings. Valves FCV-104-17A/B/C/D and E are normally closed valves that are only open when the traveling screens are being washed. Providing water to clean the screens is not a function that meets the requirements of 10 CFR 54.4(a)(1) or (a)(3). These valves fail to a closed position such that failure of the piping downstream of these valves would not affect the ability of the SW system to perform its functions as required by 10 CFR 54.4(a)(1) or (a)(3). However, as described in LRA Table 2.3.3.13-B, the portion of the SW system in the intake structure near the SW pumps and the components outside the safety class pressure boundary, while relied upon to provide structural/seismic support for the pressure boundary are in-scope and subject to an AMR as required by 10 CFR 54.4(a)(2). This includes the portion of lines downstream of FCV-104-17A/B/C/D and E that provide structural support.

Based on its review, the staff found the applicant response to RAI 2.3.3.2a-3 acceptable because the applicant acknowledged these sections of piping are within the scope of license renewal. As described in LRA Section 2.1.2.1.3, portions of systems included for 10 CFR 54.4(a)(2) are not shown on LRA drawings. Although the applicant did not identify these sections of piping as being within the boundary of license renewal on the drawing, the applicant confirmed they are within the scope based on the potential for physical interaction with safety-related systems in accordance with 10 CFR 54.4(a)(2). Therefore, the staff concern described in RAI 2.3.3.2a-3 is resolved.

The staff noted that license renewal drawing LRA-G-191159-SH-02-0, at location G-6, depicts a license renewal boundary flag at the tee of pipe sections 2"-SW-566D and 8"-SW-34. There are no highlighted pipes or components on 2"-SW-566D or 8"-SW-34. In RAI 2.3.3.2a-4 dated August 16, 2006, the staff requested that the applicant clarify which portions of pipe and components are and are not bounded by the aforementioned boundary flag and justify the boundary locations with respect to the applicable requirements of 10 CFR 54.4(a).

In its response dated September 20, 2006, the applicant stated license renewal drawings only show the portions of the system with intended functions that meet the requirements of 10 CFR 54.4(a)(1) or (a)(3). As described in LRA Section 2.1.2.1.3, portions of systems included as required by 10 CFR 54.4(a)(2) are not shown on license renewal drawings. The piping and valves on line 2"-SW- 566D are safety-related, since they have a safety function to break vacuum and prevent water hammer in the SW system. As a result, a system intended function boundary flag is provided that points towards and includes all the components on line 2"-SW-566D. The reason these components are not highlighted as subject to an AMR is that they perform their system intended function though the active function of the valves opening and breaking vacuum. In accordance with 10 CFR 54.21 (a)(1)(i), components that perform their intended functions with moving parts or a change in configuration are not subject to an AMR. These components do not have a passive intended function of pressure boundary as required by 10 CFR 54.4(a)(1) or (a)(3), since this portion of the system is isolated when aligned to the ultimate heat sink. However, as described in LRA Table 2.3.3.13-6, the portion of the SW system inside the reactor building and the components outside the safety class pressure boundary, while relied upon to provide structural/seismic support for the pressure boundary are in-scope and subject to an AMR as required by 10 CFR 54.4(a)(2). This includes line 2-SW-566D and portions of lines connected to this line that provide structural support and have the potential to affect safety-related components due to spray or leakage.

Based on its review, the staff found the applicant response acceptable because the applicant acknowledged that pipe 2" SW-566D is within the scope of license renewal and subject to an AMR based on the potential for physical interaction with safety-related systems in accordance with 10 CFR 54.4(a)(2). As described in LRA Section 2.1.2.1.3, portions of systems included for 10 CFR 54.4(a)(2) are not shown on LRA drawings. Therefore, the staff concern described in RAI 2.3.3.2a-4 is resolved.

The staff's review of LRA Section 2.3.3.2 identified areas in which information provided in the LRA needed to be confirmed by the NRC Regional Inspection Team to complete the review of the applicant's scoping and screening results.

Inspection Item 2.3.3.2a-1

License renewal drawing LRA-G-191159-SH-01-0, at location H-11, depicts pipe section 2"-SW-566C as within the scope of license renewal. The license renewal boundary flag for 2"-SW-566C is located on an unisolable section of pipe. The actual location of the license renewal scope boundary for this pipe section is not clear. The staff requested that the NRC Regional Inspection Team perform an inspection to ensure that the license renewal scope boundaries for these components meet the requirements of 10 CFR 54.4(a)(2). The staff identified this as confirmatory item 2.3.3.2a-1.

In Inspection Report 05000271/2007006, Vermont Yankee Nuclear Power Station - NRC License Renewal Inspection Report, dated June 4, 2007, Attachment, Review of Safety Evaluation Report Confirmatory Items, the regional inspection team stated in part that the applicant has included in-scope for spatial interaction the portion of the SW system in the service water pump area of the intake structure and the reactor building. Pipe section 2" SW-566C is in the reactor building and is therefore in-scope for spatial interaction. As described in LRA Section 2.1.2.1.3, portions of systems included for 10 CFR 54.4(a)(2) are not shown on LRA drawings. Further, the applicant's letter to the NRC dated July 3, 2007, LRA Amendment 27, Attachment 2 indicates that pipe section 4"SW-567 which attaches to pipe section 2" SW-566C is in-scope for spatial interaction.

Based on its review, the staff found the above response acceptable because the inspection team and the applicant acknowledged that service water pipe 2" SW-566C is within the scope of license renewal and subject to an AMR based on the potential for physical interaction with safety-related systems in accordance with 10 CFR 54.4(a)(2). Therefore, the staff concern described in confirmatory item 2.3.3.2a-1 is resolved.

Inspection Item 2.3.3.2a-2

LRA Section 2.1.2.1.2 states in part that nonsafety-related piping systems connected to safety-related systems were included up to the structural boundary or to a point that includes an adequate portion of the nonsafety-related piping run to conservatively include the first seismic or equivalent anchor. In addition, if isometric drawings were not readily available to identify the structural boundary, connected lines were included to a point beyond the safety/nonsafety interface, like a base-mounted component, flexible connection, or the end of a piping run (*i.e*, a drain line).

The staff cannot determine whether all the nonsafety-related piping systems were included up to the structural boundary or to a point that includes an adequate portion of the nonsafety-related piping run to include the first seismic or equivalent anchor. The staff requested that the NRC Regional Inspection Team perform an inspection to ensure that the license renewal scope boundaries for these components satisfy the requirements of 10 CFR 54.4(a)(2). The staff identified this as confirmatory item 2.3.3.2a-2.

In Inspection Report 05000271/2007006, Vermont Yankee Nuclear Power Station - NRC License Renewal Inspection Report, dated June 4, 2007, Attachment, Review of Safety Evaluation Report Confirmatory Items, the NRC Regional Inspection Team stated in part that for structural support considerations, the applicant has included components outside the safety class pressure boundary, yet relied upon to provide structural/seismic support for the pressure boundary. The application describes the types of components which are included in the scope of license renewal for 10 CFR 54.4(a)(2) and subject to an AMR in the service water system in LRA Table 2.3.3-13-42. This table was developed by including all nonsafety-related portions of fluid systems which are located within a building containing safety-related components and all nonsafety-related piping connected to safety-related systems back to the structural boundary using an isometric drawing. In cases where an isometric drawing which depicts the structural boundary is not readily available, connected lines were included back to a point beyond the safety/nonsafety interface to a base-mounted component, flexible connection, or the end of a piping run (such as a drain line) in accordance with the response to RAI 2.1-2. As described in LRA Section 2.1.2.1.3, portions of systems included for 10 CFR 54.4(a)(2) are not shown on LRA drawings.

Further, the applicant's letter to the NRC dated July 3, 2007, LRA Amendment 27, Attachment 2 states that there are no nonsafety-related systems for which the applicant has not identified the nonsafety-related portions of systems which are attached to safety-related systems and required to be in the scope of license renewal in accordance with 10 CFR 54.4(a)(2). However, as a result of discussions with the staff during the Region I inspection (February 2007), the applicant determined that some safety-related SSCs in the VY turbine building required consideration for potential spatial impacts from nonsafety-related SSCs based on 10 CFR 54.4(a)(2). Therefore, an expanded review for SSCs in the turbine building determined that additional components required an AMR. Those additional component types have been added to LRA Table 2.3.3-13-42, as addressed in the applicant's letters to the NRC dated July 30, 2007 and August 16, 2007.

Based on its review, the staff finds the response acceptable because the NRC Regional Inspection Team found there are no nonsafety-related portions of systems which are attached to safety-related systems that are not within the scope of license renewal in accordance with 10 CFR 54.4(a)(2). Furthermore, the staff again reviewed the applicable LRA drawings for component types that may have been omitted from Table 2.3.3-13-42 and found all component types in Table 2.3.3-13-42 to be consistent with the component types included within the scope of license renewal at similar facilities. Therefore, the staff concern described in confirmatory item 2.3.3.2a-2 is resolved.

2.3.3.2.3 Conclusion

The staff reviewed the LRA, accompanying license renewal drawings, and RAI and confirmatory item responses to determine whether the applicant failed to identify any SSCs within the scope of license renewal or subject to an AMR. The staff finds no such omissions. On the basis of its review, the staff concludes that there is reasonable assurance that the applicant has adequately identified the SW and RHRSW system components within the scope of license renewal, as required by 10 CFR 54.4(a), and those subject to an AMR, as required by 10 CFR 54.21(a)(1).

2.3.3.3 Reactor Building Closed Cooling Water

2.3.3.3.1 Summary of Technical Information in the Application

LRA Section 2.3.3.3 describes the RBCCW system, which supplies demineralized water to the reactor building auxiliary equipment systems from a closed cooling loop. The RBCCW system cools equipment which may contain radioactive fluids. The SW system provides the heat sink for the RBCCW system. The RBCCW cooling function is not a safety function. FPC is not a safety function of RBCCW since the safety-related SBFPC system uses SW as a heat sink. RBCCW supplies the heat sink for the nonsafety-related FPC system. RHR pump seal cooling is normally provided by RBCCW, not SW. This is not a safety function for RBCCW because RHR pump seal cooling is not required to support hot safe shutdown. However, if the SW pumps are inoperable and alternate cooling is inservice, the RHR pump seal coolers are manually aligned to the SW supplied by the ACS. In accordance with these conditions (loss of Vernon Pond, flooding of the SW intake structure, or fire in the SW intake structure which disables all four SW pumps), RHR pump seal cooling is a safety function of SW via ACS and the RBCCW system piping, which provides for seal cooling to be supplied by ACS and performs the safety function of maintaining SW system integrity.

The RBCCW system has safety-related components relied upon to remain functional during and following DBEs. The failure of nonsafety-related SSCs in the RBCCW system potentially could prevent the satisfactory accomplishment of a safety-related function. In addition, the RBCCW system performs functions that support fire protection.

LRA Tables 2.3.3-3 and 2.3.3-13-30 identify the following RBCCW system component types within the scope of license renewal and subject to an AMR:

*   bolting
*   flow switch housing
*   heat exchanger (housing)
*   heat exchanger (shell)
*   heat exchanger (tubes)
*   piping
*   pump casing
*   sight glass
*   strainer housing

- tank
- thermowell
- tubing
- valve body

The RBCCW system component intended functions within the scope of license renewal include the following:

- pressure boundary
- structural or functional support for safety-related equipment

2.3.3.3.2 Staff Evaluation

The staff reviewed LRA Section 2.3.3.3 and UFSAR Section 10.9 using the Tier-2 evaluation methodology described in SER Section 2.3 and the guidance in SRP-LR Section 2.3.

In conducting its review, the staff evaluated the system functions described in the LRA and UFSAR to verify that the applicant has not omitted from the scope of license renewal any components with intended functions as required by 10 CFR 54.4(a). The staff then reviewed those components that the applicant has identified as within the scope of license renewal to verify that the applicant has not omitted any passive and long-lived components subject to an AMR as required by 10 CFR 54.21(a)(1).

The staff's review of LRA Section 2.3.3.3 identified an area in which additional information was necessary to complete the review of the applicant's scoping and screening results. The applicant responded to the staff's RAI as discussed below.

The staff noted that license renewal drawing LRA-G-191159-SH-03-0, at location P-10 at valve 29 shows a section of pipe within the scope of license renewal. This section of pipe is the RBCCW return to the ACS. However, a drawing continuation is not provided. In RAI 2.3.3.3-1 dated August 16, 2006, the staff requested that the applicant provide information for the continuation of this piping section to the license renewal boundary and justify the boundary location with respect to the applicable requirements of 10 CFR 54.4(a).

In its response dated September 20, 2006, the applicant stated that the RBCCW return to the ACS shown on license renewal drawing LRA-G-191159-SH-03-0, at location P-10 at valve 29 continues on license renewal drawing LRA-G-191159-SH-02-0, at location E-2.

Based on its review, the staff found the applicant response to RAI 2.3.3.3-1 acceptable because the applicant provided the necessary drawings and documentation to demonstrate this section of reactor building closed cooling water piping was connected to the service water system, was identified as being within the scope of license renewal, and with boundaries correctly identified on the service water system flow diagram, LRA-G-191159-SH–2-0. Therefore, the staff concern described in RAI 2.3.3.3-1 is resolved.

2.3.3.3.3 Conclusion

The staff reviewed the LRA, accompanying license renewal drawings, and RAI responses to determine whether the applicant failed to identify any SSCs within the scope of license renewal or subject to an AMR. The staff finds no such omissions. On the basis of its review, the staff concludes that there is reasonable assurance that the applicant has adequately identified the RBCCW system components within the scope of license renewal, as required by 10 CFR 54.4(a), and those subject to an AMR, as required by 10 CFR 54.21(a)(1).

2.3.3.4 Emergency Diesel Generator

2.3.3.4.1 Summary of Technical Information in the Application

LRA Section 2.3.3.4 describes the EDG and the diesel lube oil (DLO) systems. The purpose of the DG system is to provide Class 1E electrical power to the emergency buses in a loss of normal power condition or a LOCA coincident with loss of normal power or degraded grid voltage at the emergency buses and is available to provide Class 1E electrical power to the emergency buses in a LOCA with normal power available. The DG and auxiliary systems will start and be in standby during a LOCA. The purpose of the DLO system is to provide for DLO storage and provide for prelube of the DGs. The DLO system consists of two lube oil day tanks and pre-lube oil pumps only. The DLO system in the component database has only these four components. The remaining components supplying lube oil required during EDG operation are in the DG system.

The DG and DLO systems have safety-related components relied upon to remain functional during and following DBEs. The failure of nonsafety-related SSCs in the system potentially could prevent the satisfactory accomplishment of a safety-related function. In addition, the systems perform functions that support fire protection.

LRA Tables 2.3.3-4, 2.3.3-13-10, and 2.3.3-13-11 identify the following EDG system, DG and auxiliaries system, and DLO system component types within the scope of license renewal and subject to an AMR:

- bolting
- expansion joint
- filter housing
- heat exchanger (bonnet)
- heat exchanger (fins)
- heat exchanger (shell)
- heat exchanger (tubes)
- heat exchanger (tubesheets)
- heater housing
- orifice
- piping
- pump casing
- sight glass
- silencer
- strainer

2-70

- strainer housing
- tank
- thermowell
- tubing
- turbocharger
- valve body

The component intended functions within the scope of license renewal include the following:

- flow control
- filtration
- heat transfer
- pressure boundary

2.3.3.4.2 Staff Evaluation

The staff reviewed LRA Sections 2.3.3.4 and 2.3.3.13, and UFSAR Section 8.5 using the Tier-2 evaluation methodology described in SER Section 2.3 and the guidance in SRP-LR Section 2.3.

In conducting its review, the staff evaluated the system functions described in the LRA and UFSAR to verify that the applicant has not omitted from the scope of license renewal any components with intended functions as required by 10 CFR 54.4(a). The staff then reviewed those components that the applicant has identified as within the scope of license renewal to verify that the applicant has not omitted any passive and long-lived components subject to an AMR as required by 10 CFR 54.21(a)(1).

In letters to the NRC dated July 30, 2007 and August 16, 2007, the applicant reported the deletion of DG compressor housing from LRA Table 2.3.3-13-10 as a component type subject to an AMR. The applicant stated that since the compressor housing will not contain liquid, it should not be subject to an AMR for potential spatial interaction. The staff has reviewed this component type deletion and concurs that the deletion of the DG compressor housing is acceptable.

2.3.3.4.3 Conclusion

The staff reviewed the LRA, accompanying license renewal drawings, and RAI responses to determine whether the applicant failed to identify any SSCs within the scope of license renewal or subject to an AMR. The staff finds no such omissions. On the basis of its review, the staff concludes that there is reasonable assurance that the applicant has adequately identified the EDG system, DG and auxiliaries system, and DLO system components within the scope of license renewal, as required by 10 CFR 54.4(a), and those subject to an AMR, as required by 10 CFR 54.21(a)(1).

2.3.3.5 Fuel Pool Cooling

2.3.3.5.1 Summary of Technical Information in the Application

LRA Section 2.3.3.5 describes the FPC system, the safety-related SBFPC subsystem, the fuel pool filter-demineralizer (FPFD) system, and the Boral in the spent fuel racks. The FPC system removes the decay heat released from the spent fuel elements. During normal operation, the system maintains a specified fuel pool water temperature, purity, water clarity, and water level. The system cools the fuel storage pool by transferring the spent fuel decay heat through heat exchangers to the RBCCW. The purpose of the SBFPC system is to maintain pool temperature during design basis accidents (including concurrent LOCAs, loss of offsite power, and single failure) or if an unusually high spent fuel decay heat load is placed in the pool. The purpose of the FPFD is to maintain the purity of the spent fuel pool water by minimizing corrosion product buildup and controlling water clarity, minimizing fission product contamination in the water, and controlling removal of water from the fuel pool to the CST system. Boral sheets in the spent fuel storage pool provide neutron absorption.

The FPC and SBFPC systems have safety-related components relied upon to remain functional during and following DBEs. The failure of nonsafety-related FPC, SBFPC, and FPFD systems SSCs potentially could prevent the satisfactory accomplishment of a safety-related function. In addition, the FPC and SBFPC systems perform functions that support fire protection.

LRA Tables 2.3.3-5, 2.3.3-13-16, 2.3.3-13-17, and 2.3.3-13-37 identify the following FPC, FPFD, and SBFPC system component types within the scope of license renewal and subject to an AMR:

- bolting
- filter housing
- heat exchanger (shell)
- heat exchanger (tubes)
- neutron absorber (boral)
- orifice
- piping
- pump casing
- thermowell
- tubing
- valve body

The component intended functions within the scope of license renewal include the following:

- heat transfer
- neutron absorption
- pressure boundary

2.3.3.5.2 Staff Evaluation

The staff reviewed LRA Sections 2.3.3.5 and 2.3.3.13, and UFSAR Sections 10.3 and 10.5 using the Tier-2 evaluation methodology described in SER Section 2.3 and the guidance in SRP-LR Section 2.3.

In conducting its review, the staff evaluated the system functions described in the LRA and UFSAR to verify that the applicant has not omitted from the scope of license renewal any components with intended functions as required by 10 CFR 54.4(a). The staff then reviewed those components that the applicant has identified as within the scope of license renewal to verify that the applicant has not omitted any passive and long-lived components subject to an AMR as required by 10 CFR 54.21(a)(1).

The staff's review of LRA Section 2.3.3.5 identified an area in which additional information was necessary to complete the review of the applicant's scoping and screening results. The applicant responded to the staff's RAI as discussed below.

The staff noted that license renewal drawing G-191173, Sheet 1, at location H-5 shows a section of pipe within the scope of license renewal. The section of pipe includes check valve V-30 and a "penetration at concrete wall," with changes in seismic classifications at each end. The section of pipe is isolated from all other in-scope piping and is not in an in-scope flow path. Piping upstream of V-30 (8"-FPC-24, 6"-FPC-24, and 8"-FPC-34) contains two normally closed valves (V-28 and V-53) and is not shown within the scope of license renewal. Piping downstream of V-30 (4"-FPC-24 and 4"-FPC-25) is also not shown within the scope of license renewal. Failure of these sections of piping could have an adverse effect on the intended pressure boundary function for the FPC piping. In RAI 2.3.3.5a-1 dated August 16, 2006, the staff requested that the applicant provide information to justify exclusion from the scope of license renewal the piping from valves V-28 and V-53 to valve V-30 and from the reactor well diffusers to the current license renewal boundary at the penetration upstream of valve V-30.

In its response dated September 20, 2006, the applicant stated that license renewal drawings only show the portions of the system with intended functions that meet the requirements of 10 CFR 54.4(a)(1) or (a)(3). As described in LRA Section 2.1.2.1.3, portions of systems required by 10 CFR 54.4(a)(2) are not shown on license renewal drawings. The piping from valves V-28 and V-53 to valve V-30 and from the reactor well diffusers to the license renewal boundary at the penetration upstream of valve V-30 are within the scope of license renewal and subject to an AMR as required by 10 CFR 54.4(a)(2) and as described in LRA Table 2.3.3.13-B for the FPC system. The description includes portions of the system in the primary containment building and reactor building and components outside the safety class pressure boundary which are relied upon to provide structural/seismic support for the pressure boundary. The piping in question is inside the reactor building and attached to safety-related components so it is within the scope of license renewal and subject to an AMR.

Based on its review, the staff found the applicant response to RAI 2.3.3.5a-1 acceptable because the applicant acknowledged that piping from valves V-28 and V-53 to valve V-30 and from the reactor well diffusers to the license renewal boundary at the penetration upstream of valve V-30 are included within the scope of license renewal. As described in LRA Section 2.1.2.1.3, portions of systems included for 10 CFR 54.4(a)(2) are not shown on LRA drawings. Although the applicant did not identify these sections of piping within the boundary of license renewal on the drawing, the applicant confirmed they are within the scope of license renewal based on the potential for physical interaction with safety-related systems in accordance with 10 CFR 54.4(a)(2). Therefore, the staff concern described in RAI 2.3.3.5a-1 is resolved.

2.3.3.5.3 Conclusion

The staff reviewed the LRA, accompanying license renewal drawings, and RAI response to determine whether the applicant failed to identify any SSCs within the scope of license renewal or subject to an AMR. The staff finds no such omissions. On the basis of its review, the staff concludes that there is reasonable assurance that the applicant has adequately identified the FPC, FPFD, and SBFPC system components within the scope of license renewal, as required by 10 CFR 54.4(a), and those subject to an AMR, as required by 10 CFR 54.21(a)(1).

2.3.3.6 Fuel Oil

2.3.3.6.1 Summary of Technical Information in the Application

LRA Section 2.3.3.6 describes the fuel oil (FO) system, which supplies FO to the EDGs as well as the nonsafety-related diesel-driven fire pump, John Deere diesel (JDD), and house HB. The portion of the system related to the EDGs consists of a day tank and fuel transfer pump for each diesel, the FO storage tank, valves, and piping. The diesel fire pump FO day tank, JDD day tank, and house HB FO storage tank are not connected to the FO storage tank. Normal makeup to the house HB FO storage tank is by tanker truck. Normal makeup to the diesel fire pump FO day tank and JDD day tank is from a 500-gallon portable tank filled from the FO storage tank.

The FO system has safety-related components relied upon to remain functional during and following DBEs. The failure of nonsafety-related SSCs in the FO system potentially could prevent the satisfactory accomplishment of a safety-related function. In addition, the FO system performs functions that support fire protection.

LRA Tables 2.3.3-6 and 2.3.3-13-14 identify the following FO system component types within the scope of license renewal and subject to an AMR:

- bolting
- filter housing
- flame arrestor
- flex hose
- injector housing
- piping
- pump casing
- sight glass

- strainer housing
- tank
- thermowell
- tubing
- valve body
- strainer housing

The FO system component intended functions within the scope of license renewal include the following:

- flow control
- pressure boundary

2.3.3.6.2 Staff Evaluation

The staff reviewed LRA Sections 2.3.3.6 and 2.3.3.13, and UFSAR Section 8.5.4 using the Tier-2 evaluation methodology described in SER Section 2.3 and the guidance in SRP-LR Section 2.3.

In conducting its review, the staff evaluated the system functions described in the LRA and UFSAR to verify that the applicant has not omitted from the scope of license renewal any components with intended functions as required by 10 CFR 54.4(a). The staff then reviewed those components that the applicant has identified as within the scope of license renewal to verify that the applicant has not omitted any passive and long-lived components subject to an AMR as required by 10 CFR 54.21(a)(1).

The staff's review of LRA Section 2.3.3.6 identified an area in which additional information was necessary to complete the review of the applicant's scoping and screening results. The applicant responded to the staff's RAI as discussed below.

The staff noted that license renewal drawing LRA-G-191162, Sheet 2, provides information about the EDGs, diesel-driven fire pump, and house HB systems, supported by the FO system. However, the drawing does not provide sufficient information about the JDD system, also supported by the FO system. For example, more information is required regarding the transfer system between the 75,000-gallon FO storage tank, the day tanks for the two JDDs, and single fire pump diesel, which is necessary to provide an intended function in accordance with 10 CFR 54.4 (a)(3) in support of the fire protection regulation requirements (10 CFR 50.48). The LRA text states only that a 500-gallon portable tank is used to transport FO to the diesel day tanks. Typical components subject to an AMR for diesels like the day tank, strainer, etc., for the JDDs are not covered. In RAI 2.3.3.6-1 dated August 16, 2006, the staff requested that the applicant provide FO system drawings and describe the JDD system. The staff also requested that the applicant explain the relationship between the JDD and the FO systems and clarify what the AMR tables should include in both Sections 2.3.3.6 and 2.3.3.12. The staff further requested that the applicant also provide information for the license renewal boundary that justifies its location with respect to the applicable requirements of 10 CFR 54.4(a).

In its response dated September 20, 2006, the applicant stated that the 350-gallon diesel fire pump FO day tank and 550-gallon fiberglass underground storage tank for the JDD are filled with FO from the FO storage tank. The FO is pumped from the FO storage tank drain line into a portable 500-gallon tank. The portable tank is then moved to the intake structure or JDD building by a fork lift. A 12VDC pump on the portable tank then pumps the FO into the diesel fire pump FO day tank or the fiberglass underground storage tank for the JDD. Since the portable tank and pump are not part of the FO system pressure boundary and since levels in the diesel fire pump FO day tank and underground storage tank for the JDD are maintained, the portable tank and pump do not perform a component intended function and are not subject to an AMR. A dedicated 550-gallon fiberglass underground storage tank provides fuel to the JDD engine. As the JDD is required for compliance with the staff's regulations concerning fire protection (10 CFR 50.48), providing FO for the engine is an intended function of the FO system in accordance with 10 CFR 54.4 (a)(3). Therefore, the storage tank and associated piping and components that supply FO to the diesel engine injectors are within the scope of license renewal and subject to an AMR. JDD FO components are included in LRA Tables 2.3.3.6 and 3.3.2-6. As the JDD is required for compliance with the staff's regulations concerning fire protection (10 CFR 50.48), it is within the scope of license renewal and subject to an AMR in accordance with 10 CFR 54.4 (a)(3). The passive mechanical components of the diesel subject to an AMR that were confirmed by walkdown are included in LRA Tables 2.3.3-12 and 3.3.2-12.

Based on its review, the staff found the applicant response to RAI 2.3.3.6-1 acceptable because the applicant explained that the 550-gal fiberglass underground storage tank and associated piping and components that supply FO to the diesel engine injectors are within the scope of license renewal and an AMR. The applicant stated that flow diagrams are not available for this skid-mounted diesel, or its FO system, and only a few components are represented in the equipment database. The applicant, however, has verified by walkdown of the system that these passive components are identified in AMR Tables 2.3.3-12 and 3.3.2-12. Therefore, the staff concern described in RAI 2.3.3.6-1 is resolved.

2.3.3.6.3 Conclusion

The staff reviewed the LRA, accompanying license renewal drawings, and RAI response to determine whether the applicant failed to identify any SSCs within the scope of license renewal or subject to an AMR. The staff finds no such omissions. On the basis of its review, the staff concludes that there is reasonable assurance that the applicant has adequately identified the FO system components within the scope of license renewal, as required by 10 CFR 54.4(a), and those subject to an AMR, as required by 10 CFR 54.21(a)(1).

2.3.3.7 Instrument Air

2.3.3.7.1 Summary of Technical Information in the Application

LRA Section 2.3.3.7 describes the IA, SA, 105 (IA and SA instruments), and nitrogen (N_2) supply systems. The purpose of the IA system is to provide the station continuously with dry, oil-free air for pneumatic instruments and controls through a dual header system. The IA system includes the containment N_2 supply described in the UFSAR as a separate N_2 subsystem also known as containment air. The purpose of containment N_2 is to provide pneumatically-operated components in the drywell with N_2 when the primary containment is inerted so any component leakage will not dilute the N_2 atmosphere. This N_2 source can be from either the N_2 system (normal supply) or the containment air compressor (automatic backup supply). When neither N_2 supply is available or when the containment is not inerted, IA may be lined up manually as a secondary backup for the containment N_2. When the containment is not inerted, IA will be lined up as the primary source of pneumatic pressure.

The purpose of the SA system is to provide the station with the compressed air requirements for pneumatic instruments and controls and general station services. The IA system also supports this function. The purpose of the 105 system is to provide indication, alarm, and control functions for associated systems. This code is used in the component database for various instrumentation components related to IA and SA. Although the 105 system consists mainly of EIC components, certain IA instrumentation mechanical components are included as well. The purpose of the N_2 system is to provide N_2 gas to the primary containment atmospheric control (PCAC) system to satisfy the primary containment purge and normal make-up requirements.

The IA, SA, 105, and N_2 systems have safety-related components relied upon to remain functional during and following DBEs. The failure of nonsafety-related SSCs in the IA and N_2 system potentially could prevent the satisfactory accomplishment of a safety-related function. In addition, the IA system performs functions that support fire protection and SBO.

LRA Tables 2.3.3-7, 2.3.3-13-54, 2.3.3-13-22, and 2.3.3-13-24 identify the following IA, SA and N_2 system component types within the scope of license renewal and subject to an AMR:

- bolting
- piping
- strainer housing
- tank
- trap
- tubing
- valve body

The IA, SA and N_2 system component intended function within the scope of license renewal is to provide a pressure boundary.

2.3.3.7.2 Staff Evaluation

The staff reviewed LRA Sections 2.3.3.7 and 2.3.3.13, and UFSAR Section 10.14 using the Tier-2 evaluation methodology, for IA and N_2, and the Tier-1 methodology, for SA and 105 systems, described in SER Section 2.3 and the guidance in SRP-LR Section 2.3.

In conducting its review, the staff evaluated the system functions described in the LRA and UFSAR to verify that the applicant has not omitted from the scope of license renewal any components with intended functions as required by 10 CFR 54.4(a). The staff then reviewed those components that the applicant has identified as within the scope of license renewal to verify that the applicant has not omitted any passive and long-lived components subject to an AMR as required by 10 CFR 54.21(a)(1).

In letters to the NRC dated July 30, 2007 and August 16, 2007, the applicant reported the deletion of IA compressor housing from LRA Table 2.3.3-13-22 as a component type subject to an AMR. The applicant stated that since the compressor housing will not contain liquid, it should not be subject to an AMR for potential spatial interaction. The staff has reviewed this component type deletion and concurs that the deletion of the IA compressor housing is acceptable.

2.3.3.7.3 Conclusion

The staff reviewed the LRA and accompanying license renewal drawings to determine whether the applicant failed to identify any SSCs within the scope of license renewal or subject to an AMR. The staff finds no such omissions. On the basis of its review, the staff concludes that there is reasonable assurance that the applicant has adequately identified the IA, SA, 105, and N_2 systems components within the scope of license renewal, as required by 10 CFR 54.4(a), and those subject to an AMR, as required by 10 CFR 54.21(a)(1).

2.3.3.8 Fire Protection-Water

2.3.3.8.1 Summary of Technical Information in the Application

LRA Section 2.3.3.8 describes the fire protection-water system. The fire protection system provides fire protection for the station through the use of water, CO_2, dry chemicals, foam, detection and alarm systems, and rated fire barriers, doors, and dampers. Water for the fire protection system is from two vertical turbine-type pumps, one electric motor-driven and one diesel-driven. The pumps and drivers located in the intake structure discharge to an underground piping system serving the exterior and interior fire protection systems. The pressure in the system is maintained at approximately 100 psig by an interconnection to the SW system. A check valve in the connecting pipe prevents backflow. Through an interconnecting valve, the SW system can provide water to fire protection components in the unlikely event that both fire protection pumps are unavailable.

The failure of nonsafety-related SSCs in the fire protection-water system potentially could prevent the satisfactory accomplishment of a safety-related function. The fire protection-water system also performs functions that support fire protection.

LRA Tables 2.3.3-8 and 2.3.3-13-15 identify the following fire protection-water system component types within the scope of license renewal and subject to an AMR:

- bolting
- expansion joint
- filter
- filter housing
- flow nozzle
- gear box
- heat exchanger (bonnet)
- heat exchanger (shell)
- heat exchanger (tubes)
- heater housing

- nozzle
- orifice
- piping
- pump casing
- silencer
- strainer
- strainer housing
- tank
- tubing
- turbocharger
- valve body

In LRA Table 3.3.2-8, the applicant provides the results of the AMR.

The fire protection-water system component intended functions within the scope of license renewal include the following:

- flow control
- filtration
- heat transfer
- pressure boundary

2.3.3.8.2 Staff Evaluation

The staff reviewed LRA Sections 2.3.3.8 and 2.3.3.13, and UFSAR Section 10.11 using the evaluation methodology described in SER Section 2.3 and the guidance in SRP-LR Section 2.3.

The staff evaluated the system functions described in the LRA and UFSAR to verify that the applicant has not omitted from the scope of license renewal any components with intended functions as required by 10 CFR 54.4(a). The staff then reviewed those components that the applicant has identified as within the scope of license renewal to verify that the applicant has not omitted any passive and long-lived components subject to an AMR as required by 10 CFR 54.21(a)(1).

The staff also reviewed the VYNPS fire protection SER, dated January 13, 1978, and supplemental SERs listed in the VYNPS Facility Operating License condition 3.F. These reports are referenced in the VYNPS fire protection CLB and summarize the fire protection program and commitments required by 10 CFR 50.48 using BTP Auxiliary and Power Conversion Systems Branch (APCSB) 9.5-1, "Guidelines for Fire Protection for Nuclear Power Plants," May 1, 1976, and Appendix A to BTP APCSB 9.5-1, August 23, 1976. The staff then reviewed those components that the applicant identified as being within the scope of license renewal to verify that the applicant did not omit any passive and long-lived components that should be subject to an AMR as required by 10 CFR 54.21(a)(1).

The staff's review of LRA Section 2.3.3.8 identified areas requiring additional information necessary to complete the review of the applicant's scoping and screening results. The applicant responded to the staff's RAIs as discussed below.

In RAI 2.3.3.8-1, dated August 15, 2006, the staff stated that LRA drawing LRA-G-191163-SH-02-0, "Fire Protection System Outer Loop," shows the yard fire hydrants as out of scope (i.e., not colored in purple). The staff requested that the applicant verify whether the yard fire hydrants are in-scope of license renewal in accordance with 10 CFR 54.4(a) and subject to an AMR in accordance with 10 CFR 54.21(a)(1). If they are excluded from the scope of license renewal and not subject to an AMR, the staff requested that the applicant provide justification for the exclusion.

In its response, by letter dated September 20, 2006, the applicant stated:

> LRA drawing LRA-G-191163-SH-02-0, "Fire Protection System Outer Loop" shows that the yard fire hydrants are not subject to an AMR since they are not highlighted.
>
> As described in LRA Section 2.3.3.8:
>
> The fire protection–water system has no intended functions as required by 10 CFR 54.4(a)(1).
>
> The fire protection–water system intended functions as required by 10 CFR 54.4(a)(2) include the following:
>
> - Maintain integrity of nonsafety-related components such that no physical interaction with safety-related components could prevent satisfactory accomplishment of a safety function.
>
> The fire protection–water system intended functions as required by 10 CFR 54.4(a)(3) include the following:
>
> - Provide the capability to extinguish fires in vital areas of the plant (10 CFR 50.48).
>
> Therefore, the fire protection system is in-scope for license renewal.
>
> The piping in the outer loop performs a component pressure boundary intended function that supports the ability of the fire protection system to extinguish fires in vital areas of the plant serviced by the inner loop. If the outer loop failed, piping that provides water to fire systems in vital areas of the plant may not perform its intended function. The yard fire hydrants are isolable from the outer loop such that their failure would not impact the support of vital areas. Yard fire hydrants are not required to extinguish fires in vital areas of the plant and their failure cannot impact safety-related components. Therefore, the yard fire hydrants perform no intended function in support of the system intended functions and are not subject to an aging management review.

In evaluating this response, the staff found that it was incomplete and that review of LRA Section 2.3.3.8 could not be completed. Yard fire hydrants are included in-scope of license and excluded from an AMR. The staff finds this contrary to the original VYNPS fire protection safety evaluation and UFSAR as the CLB. In its response, the applicant stated that the yard fire hydrants perform no intended function in support of the system intended functions and are not subject to an AMR and therefore, not credited in accordance with 10 CFR 50.48. This resulted in the staff holding a telephone conference with the applicant on November 7, 2006, to discuss information necessary to resolve the concern in RAI 2.3.3.8-1. The staff explained that the scope of SSCs required for compliance with 10 CFR 50.48 and 10 CFR 50 Appendix A, GDC 3, goes beyond preserving the ability to maintain safe-shutdown in the event of a fire. The staff stated that the exclusion of fire protection SSCs, on the basis that the intended function is not required for the protection of safe-shutdown equipment or safety-related equipment is not acceptable, if the SSC is required from compliance with 10 CFR 50.48.

By letter dated December 4, 2006, the applicant stated that the yard fire hydrants are in-scope and subject to an AMR. The hydrants are identified as component type "valve body" in LRA Table 2.3.3-8. Results of the AMR are provided in LRA Table 3.3.2-8 for line items "valve body" with carbon steel as the material and raw water as the environment.

Based on its review, the staff finds the applicant's response to RAI 2.3.3.8-1 acceptable because the applicant has committed to interpret yard fire hydrants as included in the "valve body," which is in the scope for the license renewal and subject to an AMR. The staff is adequately assured that the yard fire hydrants used for the fire suppression will be considered appropriately during the aging management activities. Therefore, the staff's concern described is RAI 2.3.3.8-1 is resolved.

In RAI 2.3.3.8-2, dated August 15, 2006, the staff stated that LRA drawing LRA-G-191163-SH-02-0, "Fire Protection System Outer Loop," shows the recirculation pump motor generator set foam system colored in purple (i.e., in-scope). This drawing does not show the 150 gallon foam concentrate tank and its components (piping and valves). The staff requested that the applicant verify whether the 150 gallon foam concentrate tank and its components are in-scope of license renewal in accordance with 10 CFR 54.4(a) and subject to an AMR in accordance with 10 CFR 54.21(a)(1). If they are excluded from the scope of license renewal and not subject to an AMR, the staff requested applicant provide justification for the exclusion.

In its response, by letter dated September 20, 2006, the applicant stated:

> LRA drawing LRA-G-191163-SH-01-0, "Fire Protection System Inner Loop" shows the recirculation pump motor generator set foam system colored in purple (i.e., subject to an AMR) at coordinates I/J-2. The associated 150 gallon foam concentrate tank (TK76-1B) and its components are in-scope and subject to an AMR as shown on the same drawing at coordinates B-8. LRA Table 3.3.2.8 includes line items for the tank and associated piping, valves, and flow nozzles with fire protection foam as the internal environment.

Based on its review, the staff found the applicant's response to RAI 2.3.3.8-2 acceptable because the recirculation pump motor generator set foam system and the 150 gallon foam concentrate tank and its components (piping and valves) were identified to be in the scope of license renewal and subject to an AMR. Therefore, the staff concludes that this recirculation pump motor generator set foam system and the associated components are correctly included in the scope of license renewal and subject to an AMR. The staff's concern described in RAI 2.3.3.8-2 is resolved.

In RAI 2.3.3.8-3, dated August 15, 2006, the staff stated that NRC SE Section 3.2.2, dated January 13, 1978, approving the VYNPS fire protection program, discusses the use of flame retardant coating to protect electrical cables in trays and risers in the switchgear room to meet the requirements of 10 CFR 50.48. The LRA does not list flame retardant coating for cables. The staff requested that the applicant verify whether the flame retardant coating is in-scope of license renewal in accordance with 10 CFR 54.4(a) and subject to an AMR in accordance with 10 CFR 54.21(a)(1). If flame retardant coating is excluded from the scope of license renewal and not subject to an AMR, the staff requested applicant provide justification for the exclusion.

In its response, by letter dated September 20, 2006, the applicant stated:

> Flame retardant (flamemastic) coatings are in-scope and subject to an AMR and are included in the line item "Fire wrap" in LRA Tables 2.4-6 and 3.5.2-6. Flamemastic was inadvertently omitted from the list of materials for the line item "Fire wrap" in LRA Table 3.5.2-6.

Based on its review, the staff found the applicant's response to RAI 2.3.3.8-3 acceptable because the applicant states that the fire retardant coating "Flamemastic" was inadvertently omitted from the list of materials for the line item "Fire wrap" in LRA Table 3.5.2-6. Because the applicant has committed to interpret fire retardant coating as included in the line item "Fire wrap," which is in the scope for license renewal and subject to an AMR, the staff is adequately assured that the fire retardant coating used to protect electrical cables in trays and risers will be considered appropriately during plant aging management activities. Therefore, the staff's concern described in RAI 2.3.3.8-3 is resolved.

In RAI 2.3.3.8-4, dated August 15, 2006, the staff stated that VYNPS fire protection safety evaluation Section 4.3.1(f) discusses a manually-operated foam maker with a permanent storage tank with fire suppression functions in the event of a fire affecting the 75,000 gallon outdoor FO storage tank, the diesel generator day tanks, or the diesel generator room located on the ground floor of the turbine building. The LRA does not list this foam maker and its associated storage tank systems and components. The staff requested that the applicant verify whether the foam maker and storage tank system and components (piping and valves) are in-scope of license renewal in accordance with 10 CFR 54.4(a) and subject to an AMR in accordance with 10 CFR 54.21(a)(1). If they are excluded from the scope of license renewal and not subject to an AMR, the staff requested applicant provide justification for the exclusion.

In its response, by letter dated September 20, 2006, the applicant stated:

> As discussed in LRA Section 2.3.3.8, in the turbine building, in addition to hose stations and deluge systems, a foam fire protection agent is available that can be used to combat fires at the FO storage tank, turbine lube oil storage tank, main and auxiliary transformers, house HBs, and the emergency diesel generators.

The turbine building foam tank (TK76-1A) and associated piping and valves are in-scope and subject to an AMR as shown on LRA drawing LRA-G-191163-SH-01-0, "Fire Protection System Inner Loop" at coordinates E-8. This manual foam system is used by attaching a fire hose to the outlet and opening valves to enable water from the fire protection header to mix with the foam concentrate from the storage tank and flow through the hose. LRA Table 3.3.2.8 includes line items for the tank and associated piping and valves with fire protection foam as the internal environment.

Fire hoses are periodically replaced and managed by the existing fire protection program, and therefore are not subject to an AMR.

Based on its review, the staff found the applicant's response to RAI 2.3.3.8-4 acceptable because the manually-operated foam maker with a permanent storage tank located on the ground floor of the turbine building was identified to be in the scope of license renewal and subject to an AMR. This foam system is to be used in the event of a 75,000 gallon outdoor FO storage tank fire, or diesel generator day tank fire, or diesel generator room fire.

Further, the applicant states that LRA Table 3.3.2.8 includes line items for the tank and associated piping and valves with fire protection foam as the internal environment. The applicant also states that the fire hoses associated with this foam system are outside the scope of license renewal since they are periodically replaced (short-lived components) and managed by the existing fire protection program. Therefore, the staff concludes that the turbine building foam systems and the associated components are correctly included in the scope of license renewal and subject to an AMR. The staff's concern described in RAI 2.3.3.8-4 is resolved.

In RAI 2.3.3.8-5, dated August 15, 2006, the staff stated that VYNPS fire protection safety evaluation Section 4.5 discusses floor drains provided in all plant areas protected with fixed water fire suppression. Are they in the scope of license renewal in accordance with 10 CFR 54.4(a) and subject to an AMR in accordance with 10 CFR 54.21(a)(1). If they are excluded from the scope of license renewal and not subject to an AMR, the staff requested applicant provide justification for the exclusion.

In its response, by letter dated September 20, 2006, the applicant stated:

> Water-filled components in the radioactive waste system (which includes the floor drain system) that could affect safety-related equipment are in-scope and require an AMR in accordance with 10 CFR 54.4(a)(2) due to potential spatial interaction. These components are subject to an AMR and are addressed in LRA Table 3.3.2-13-32.

Based on its review, the staff found the applicant's response to RAI 2.3.3.8-5 acceptable. Although the VYNPS fire protection safety evaluation addresses these floor drains as associated with fire suppression, it is not included in LRA Table 3.3.2-8 "Fire Protection–Water System." Instead, it is included in LRA Table 3.3.2-13-32, "Radwaste Liquid & Solid (RDW) Nonsafety-Related Components Affecting Safety-Related Systems," which is in the scope for license renewal and subject to an AMR. Because the applicant has committed to interpret these floor drains as included in the radioactive waste system, which is in the scope for license renewal and subject to an AMR, the staff is adequately assured that the floor drains used for fire suppression will be considered appropriately during plant aging management activities. Therefore, the staff's concern described in RAI 2.3.3.8-5 is resolved.

In RAI 2.3.3.8-6, dated August 15, 2006, the staff stated that the supplement to VYNPS fire protection safety evaluation Section 3.3, dated February 20, 1980, discusses the fire protection features for the primary containment (e.g., fixed suppression systems, standpipe and hose stations, and oil collection system). The staff requested that the applicant determine whether fire protection systems and features for primary containment should be included as systems and components in-scope for license renewal and subject to an AMR. If not, the staff requested applicant explain the basis.

In its response, by letter dated September 20, 2006, the applicant stated:

> Section 3.3 of the SE supplement dated February 20, 1980, discusses potential fire protection features for the primary containment in the event the containment is not inerted. As noted in LRA Section 3.3.2.2.7, VYNPS is a BWR with an inert containment atmosphere. Therefore, the primary containment does not have a fixed suppression system or a reactor recirculation pump oil collection system.

As shown on LRA drawing LRA-G-191163-SH-01-0, "Fire Protection System Inner Loop," hose stations in the reactor building that may be used for fire suppression in primary containment during non-inerted outage periods are in-scope and subject to an AMR.

Based on its review, the staff found the applicant's response to RAI 2.3.3.8-6 acceptable because VYNPS is a BWR with an inert containment atmosphere and the primary containment does not have a fixed suppression system or a reactor recirculation pump oil collection system. Further, the applicant states that during non-inerted outage periods, hose stations in the reactor building, may be used for fire suppression in primary containment. Therefore, the staff concludes that the fire protection features for the primary containment (e.g., fixed suppression systems, standpipe and hose stations, and oil collection system) are correctly excluded from the scope of license renewal and are not subject to an AMR. During the refueling outage, hose stations in the reactor building may be used for fire suppression in the primary containment. This system was identified to be in the scope of license renewal and subject to an AMR. Therefore, the staff's concern described in RAI 2.3.3.8-6 is resolved.

In RAI 2.3.3.8-7, dated August 15, 2006, the staff stated that the supplement to VYNPS fire protection safety evaluation Section 3.3, dated October 24, 1980, discusses the deluge system used to protect the turbine building lay-down area. The staff requested that the applicant determine whether the turbine building lay-down deluge system and its components should be included as systems and components in-scope for license renewal and subject to an AMR. If not, the staff requested applicant explain the basis.

In its response, by letter dated September 20, 2006, the applicant stated:

> The turbine building loading bay is the area referred to in the SE supplement as the turbine building lay-down area. The sprinkler system for this area is in-scope and subject to an AMR as shown on LRA drawing LRA-G-191163-SH-01-0, "Fire Protection System Inner Loop" at coordinate G-9.

Based on its review, the staff found the applicant's response to RAI 2.3.3.8-7 acceptable because the deluge system and its components were identified to be in the scope of license renewal and subject to an AMR. Therefore, the staff concludes that this turbine building lay-down area deluge system and its associated components are correctly included in the scope of license renewal and subject to an AMR. The staff's concern described in RAI 2.3.3.8-7 is resolved.

In RAI 2.3.3.8-8, dated August 15, 2006, the staff stated that VYNPS fire protection safety evaluation Section 4.3.1(e) discusses the automatic sprinkler systems used for various areas including the outdoor transformer. The LRA does not list the sprinkler systems nor associated components to protect the outdoor transformer. The staff requested that the applicant verify whether the sprinkler system and associated components are in-scope of license renewal in accordance with 10 CFR 54.4(a) and subject to an AMR in accordance with 10 CFR 54.21(a)(1). If they are excluded from the scope of license renewal and not subject to an AMR, the staff requested applicant provide justification for the exclusion.

In its response, by letter dated September 20, 2006, the applicant stated:

> As described in LRA Section 2.3.3.8, the fire protection system is in the scope of license renewal in accordance with 10 CFR 54.4(a)(3) because it is credited in the Appendix R safe-shutdown analysis as required by 10 CFR 50.48.
>
> The main transformer and auxiliary transformer sprinkler fire protection subsystems do not mitigate fires in areas containing equipment important to safe operation of the plant, nor are they credited with achieving safe-shutdown in the event of a fire. These subsystems are only required to meet state, municipal, or insurance requirements. Therefore, these subsystems have no intended function and are not included in the AMR summarized in LRA Table 3.3.2-8.
>
> Since they are outdoors and away from safety-related equipment, the main transformer and auxiliary transformer sprinkler subsystems cannot affect safety-related equipment by spatial interaction and therefore, have no intended function as required by 10 CFR 54.4(a)(2). Therefore, these subsystems are not included in the AMR summarized in LRA Table 3.3.2-13-15.

Based on its review, the staff found the applicant's response to RAI 2.3.3.8-8 acceptable. Although the main transformer and auxiliary transformer sprinkler systems are addressed in the VYNPS fire protection safety evaluation, these systems in question are not credited to meet the requirements of Appendix R for achieving safe-shutdown in the event of a fire. In addition, the staff reviewed commitments made by the applicant to satisfy Appendix A to BTP APCSB 9.5-1, which discussed that the main transformer and auxiliary transformer are either located at least

2-85

50 feet from the building containing safety-related equipment or the wall of the building is a 3-hour fire-rated wall. Therefore, the staff finds that the main transformer and auxiliary transformer cannot affect safety-related equipment by spatial interaction and the sprinkler systems for the main transformer and auxiliary transformer were correctly excluded from the scope of license renewal and not subject to an AMR. Therefore, the staff's concern described in RAI 2.3.3.8-8 is resolved.

In RAI 2.3.3.8-9, dated August 15, 2006, the staff stated that VYNPS fire protection safety evaluation Section 5.12.6 discusses the use of a 3-hour rated fire protection coating to protect the structural steel supporting the wall and ceiling of diesel generator rooms. The LRA does not list 3-hour rated fire protection coating for structural steel. The staff requested that the applicant verify whether the fire protection coating for structural steel is in-scope of license renewal in accordance with 10 CFR 54.4(a) and subject to an AMR in accordance with 10 CFR 54.21(a)(1). If fire protection coating is excluded from the scope of license renewal and not subject to an AMR, the staff requested applicant provide justification for the exclusion.

In its response, by letter dated September 20, 2006, the applicant stated:

> Subsequent to the January 17, 1978, NRC Safety Evaluation, VYNPS notified the NRC (in letter WVY 78-85) that a protective coating with a "fire resistant rating of approximately 1-hour" would be utilized for the structural steel supporting the roof and ceiling. This is based on the conclusion that a fire in one diesel generator room will not result in structural damage that could result in fire spread to the other room. The fire retardant coatings are in-scope and subject to an AMR and are included in the line item "Fire proofing" in LRA Tables 2.4-6 and 3.5.2-6.

Based on its review, the staff found the applicant's response to RAI 2.3.3.8-9 acceptable. The VYNPS fire protection safety evaluation addresses the use of a 3-hour rated fire retardant coating to protect the structural steel supporting the wall and ceiling of the diesel generator rooms. The staff has confirmed that the applicant correctly identified the actual fire resistance rating of the structural steel coating (i.e., 1 hour). The fire resistance rating of the structural steel coating was clarified and included in the LRA Tables 2.4-6 and 3.5.2-6 and the coating is within the scope of license renewal and subject to an AMR. Therefore, the staff's concern described in RAI 2.3.3.8-9 is resolved.

In RAI 2.3.3.8-10, dated August 15, 2006, the staff stated that LRA Table 2.3.3-8 excludes several types of fire protection components that appear in the VYNPS fire protection safety evaluation and its supplements and/or updated UFSAR, and which also appear in the LRA drawings colored in purple. These components are listed below.

- hose stations
- hose connections
- hose racks
- pipe fittings
- pipe supports
- couplings
- threaded connections

- flexible hoses
- restricting orifices
- interface flanges
- chamber housings
- heat-actuated devices
- gauge snubbers
- tank heaters
- thermowells
- water motor alarms
- fire hydrants (casing)
- sprinkler heads
- dikes (contain oil spill)
- flame retardant coating for cables
- fire barrier penetration seals
- fire barrier walls, ceilings, floors, and slabs
- fire doors
- fire rated enclosures
- fire retardant coating for structural steel supporting walls and ceilings

For each, the staff requested applicant determine whether the component should be included in Table 2.3.3.8, and if not, justify the exclusion.

In its response, by letter dated September 20, 2006, the applicant stated the following:

- hose stations – Since they support criterion (a)(3) equipment, hose stations are included in the structural AMR. They are included in the "Fire hose reels" line item in LRA Table 2.4-6.

- hose connections – Hose connections are included in the "Piping" line item in LRA Table 2.3.3-8.

- hose racks – Since they support criterion (a)(3) equipment, hose racks are included in the structural AMR. They are included in the "Fire hose reels" line item in LRA Table 2.4-6.

- pipe fittings – As stated in LRA Section 2.0, the term "piping" in component lists may include pipe, pipe fittings (such as elbows and reducers), flow elements, orifices, and thermowells. Pipe fittings are included in the "Piping" line item in LRA Table 2.3.3-8.

- pipe supports – Since they support criterion (a)(3) equipment, piping supports are included in the structural AMR. They are included in the "Component and piping supports" line item in LRA Table 2.4-6.

- couplings – As stated in LRA Section 2.0, the term "piping" in component lists may include pipe, pipe fittings (such as elbows and reducers), flow elements, orifices, and thermowells. Couplings are pipe fittings included in the "Piping" line item in LRA Table 2.3.3-8.

- threaded connections – As stated in LRA Section 2.0, the term "piping" in component lists may include pipe, pipe fittings (such as elbows and reducers), flow elements, orifices, and thermowells. Threaded connections are pipe fittings included in the "Piping" line item in LRA Table 2.3.3-8.

- flexible hoses – Hoses are replaced on a specified periodicity and therefore, are not subject to an AMR as required by 10 CFR 54.21(a)(1)(ii).

- restricting orifices – As stated in LRA Section 2.0, the term "piping" in component lists may include pipe, pipe fittings (such as elbows and reducers), flow elements, orifices, and thermowells. Restricting orifices are included in the "Piping" line item in LRA Table 2.3.3-8.

- interface flanges – As stated in LRA Section 2.0, the term "piping" in component lists may include pipe, pipe fittings (such as elbows and reducers), flow elements, orifices, and thermowells. Interface flanges are pipe fittings included in the "Piping" line item in LRA Table 2.3.3-8.

- chamber housings – As shown on LRA drawing LRA-G-191163-SH-01-0, the turbine building lube oil room sprinkler system includes a retard chamber, piping, and valves whose purpose is to prevent false alarms due to system pressure surges and to provide a flow path to the water gong alarm during system actuation. Since failure of these components downstream of valve DV-76-200D would not prevent fire suppression capability for the lube oil room sprinkler system, they are not subject to an AMR.

- heat-actuated devices – As stated in UFSAR Section 10.11.3, the pre-action fire protection subsystems for the hydrogen seal oil area and the turbine building condenser and heater bay area have heat-actuated devices to initiate opening of the deluge valves. Heat-actuated devices are active components; not subject to an AMR.

- gauge snubbers – Gauge snubbers are integral parts of tubing runs that protect instrumentation from pressure surges. Gauge snubbers in tubing runs to instruments are included in the "tubing" line item in LRA Table 2.3.3-8.

- tank heaters – Neither the SE and its supplements nor the UFSAR discuss tank heaters. Tank heaters do not appear on the LRA drawings colored in purple. VYNPS does not have fire water storage tanks and the foam concentrate tanks do not have heaters. Therefore, the fire protection - water system does not have tank heaters.

- thermowells – As stated in LRA Section 2.0, the term "piping" in component lists may include pipe, pipe fittings (such as elbows and reducers), flow elements, orifices, and thermowells. Thermowells are included in the "Piping" line item in LRA Table 2.3.3-8.

- water motor alarms – This response assumes that reviewer means water flow alarms which are provided in critical locations and annunciate in the control room to provide positive indication of fire water system operation. Water flow alarms are active components; not subject to an AMR.

- fire hydrants (casing) – As described in response to RAI 2.3.3.8-1, the yard fire hydrants are not subject to an AMR. By letter dated December 4, 2006, the applicant stated that the yard fire hydrants are in-scope and subject to an AMR. The hydrants are identified as component type "valve body" in LRA Table 2.3.3-8. Results of the AMR are provided in LRA Table 3.3.2-8 for line items "valve body" with carbon steel as the material and raw water as the environment.

- sprinkler heads – Sprinkler heads are included in the "Flow nozzle" line item in LRA Table 2.3.3-8.

- dikes (contain oil spill) – Dikes are included in the structural AMR. They are included in the "Flood curb" line items in LRA Table 2.4-6.

- flame retardant coating for cables – As described in response to RAI 2.3.3.8-3, flame retardant (flamemastic) coatings are subject to an AMR and are included in the line item "Fire wrap" in LRA Table 2.4-6. Flamemastic was inadvertently omitted from the list of materials for the line item "Fire wrap" in LRA Table 3.5.2-6.

- fire barrier penetration seals – Fire barrier penetration seals are included in the structural AMR. They are included in the "Penetration sealant (fire, flood, radiation)" line item in Table 2.4-6.

- fire barrier walls, ceilings, floor, and slabs – Fire barrier walls, ceilings, floor, and slabs are included in the structural AMR. They are included in the concrete line items in Tables 2.4-2 through 2.4-4.

- fire doors – Fire doors are included in the structural AMR. They are included in the "Fire doors" line item in Table 2.4-6.

- fire rated enclosures – As stated in SE Section 5.17.1, the diesel day tank for the fire pump is located in a separate 3-hour fire rated enclosure. This enclosure consists of concrete block walls in the intake structure and is included in the structural AMR. It is included in the "Masonry walls" line item in Table 2.4-3.

- fire retardant coating for structural steel supporting wall and ceiling – As described in response to RAI 2.3.3.8-9, fire retardant (flamemastic) coatings are subject to an AMR and are included in the line item "Fire wrap" in LRA Table 2.4-6. Flamemastic was inadvertently omitted from the list of materials for the line item "Fire wrap" in LRA Table 3.5.2-6.

Based on its review, the staff found the applicant's response to RAI 2.3.3.8-10 acceptable. Although the applicant states that they consider these components to be included in other line items, the descriptions of the line items in the LRA do not list all these components specifically. The applicant properly identified the following components to be included in the other line items in the scope of license renewal and subject to an AMR: hose racks, pipe fittings, pipe supports, couplings, threaded connections, restricting orifices, interface flanges, gauge snubbers, thermowells, sprinkler heads, dikes, flame retardant coating for cables, fire barrier penetration seals, fire barrier walls, ceilings, floors, slabs, fire doors, fire rated enclosures, and fire retardant coating for structural steel supporting walls and ceilings. The staff is adequately assured that these components will be considered appropriately during the plant aging management activities. For each of the following components, the staff found that they were not included in the line item descriptions in the LRA: flexible hoses, chamber housings, heat-actuated devices, tank heaters, and water motor alarms. The staff recognizes the applicant's interpretation of these components as active or short-lived components will result in more vigorous oversight of the condition and performance of the components. Because the applicant has interpreted that these components are active, the staff concludes that the components were correctly excluded from the scope of license renewal and are not subject to an AMR. Therefore, the staff's concern described in RAI 2.3.3.8-10 is resolved.

In RAI 2.3.3.8-11, dated August 15, 2006, the staff stated that LRA Table 2.3.3-8 listed flow nozzles (flow control) as in-scope and subject to an AMR, but does not list spray nozzles (water). The staff requested applicant to explain why the water spray nozzles are not subject to an AMR.

In its response, by letter dated September 20, 2006, the applicant stated:

Water spray nozzles are in-scope and subject to an AMR. They are included in the line item "Flow nozzles" in LRA Table 2.3.3-8.

Based on its review, the staff finds the applicant's response to RAI 2.3.3.8-11 acceptable because it adequately explains that the spray nozzles in question are within the scope of license renewal and subject to an AMR. Further, the applicant stated that the spray nozzles are represented in the LRA Table by the component type "Flow nozzles" in LRA Table 2.3.3-8." Therefore, the staff's concern described in RAI 2.3.3.8-11 is resolved.

2.3.3.8.3 Conclusion

The staff reviewed the LRA to determine whether the applicant failed to identify any SSCs within the scope of license renewal or subject to an AMR. The staff finds no such omissions. On the basis of its review, the staff concludes that there is reasonable assurance that the applicant has adequately identified the fire protection-water system components within the scope of license renewal, as required by 10 CFR 54.4(a), and those subject to an AMR, as required by 10 CFR 54.21(a)(1).

2.3.3.9 Fire Protection-Carbon Dioxide

2.3.3.9.1 Summary of Technical Information in the Application

LRA Section 2.3.3.9 describes the fire protection-CO_2 system. The purpose of the fire protection system is to provide fire protection for the station through the use of water, CO_2, dry chemicals, foam, detection and alarm systems, and rated fire barriers, doors, and dampers. The cable vault and switchgear rooms are protected by fully automatic total flooding CO_2 suppression systems initiated by ionization detectors. Bottles located in the west switchgear room also may provide a backup or second shot to the cable vault if desired. The diesel fire pump FO storage tank room is protected by a total flooding CO_2 suppression system initiated by heat detectors. The automatic total flooding high-pressure CO_2 gas suppression systems for the cable vault and diesel fire pump FO storage tank room store high-pressure CO_2 at ambient temperatures in steel CO_2 tanks. Empty fixed piping systems convey CO_2 from the tanks to open nozzles in the fire area. The cable vault CO_2 system (automatic total flooding system with CO_2 tanks in the cable vault) is cross-connected to the CO_2 tanks in the west switchgear room for back-up capability for cable vault fire protection. The east and west switchgear rooms are protected by automatic total flooding low-pressure CO_2 systems. Low-pressure CO_2 is stored at approximately 0 °F in an outside storage tank. Empty fixed piping systems convey CO_2 from the storage tank to open nozzles in the fire area.

The fire protection-CO_2 system performs functions that support fire protection.

LRA Table 2.3.3-9 identifies the following fire protection-CO_2 system component types within the scope of license renewal and subject to an AMR:

- bolting
- coil
- filter housing
- heater housing
- nozzle
- orifice
- piping
- pump oaoing
- siren body
- strainer
- tank
- tubing
- valve body

In LRA Table 3.3.2-9, the applicant provides the results of the AMR.

The fire protection-CO_2 system component intended functions within the scope of license renewal include the following:

- flow control
- filtration
- pressure boundary

2.3.3.9.2 Staff Evaluation

The staff reviewed LRA Section 2.3.3.9 and UFSAR Section 10.11 using the evaluation methodology described in SER Section 2.3 and the guidance in SRP-LR Section 2.3.

The staff evaluated the system functions described in the LRA and UFSAR to verify that the applicant has not omitted from the scope of license renewal any components with intended functions as required by 10 CFR 54.4(a). The staff then reviewed those components that the applicant has identified as within the scope of license renewal to verify that the applicant has not omitted any passive and long-lived components subject to an AMR as required by 10 CFR 54.21(a)(1).

The staff also reviewed the approved fire protection SER, dated January 13, 1978, approving the VYNPS fire protection program and supplemental SERs listed in the VYNPS Facility Operating License condition 3.F. This report is referenced directly in the VYNPS fire protection CLB and summarizes the fire protection program and commitments to requirements of 10 CFR 50.48 using BTP APCSB 9.5-1, "Guidelines for Fire Protection for Nuclear Power Plants," May 1, 1976, and Appendix A to BTP APCSB 9.5-1, August 23, 1976. The staff then reviewed those components that the applicant identified as being within the scope of license renewal to verify that the applicant did not omit any passive and long-lived components that should be subject to an AMR as required by 10 CFR 54.21(a)(1).

The staff's review of LRA Section 2.3.3.9 identified areas requiring additional information necessary to complete the review of the applicant's scoping and screening results. The applicant responded to the staff's RAIs as discussed below.

In RAI 2.3.3.9-1, dated August 15, 2006, the staff stated that VYNPS fire protection safety evaluation Sections 3.1.5 and 4.3.2 discuss a total flooding CO_2 system for the cable spreading area, battery room, and diesel driven fire water pump tank room. The LRA does not list the CO_2 system for the cable spreading area, battery room, and diesel driven fire water pump tank room. The staff requested that the applicant verify whether the CO_2 system and its components are in-scope of license renewal in accordance with 10 CFR 54.4(a) and subject to an AMR in accordance with 10 CFR 54.21(a)(1). If they are excluded from the scope of license renewal and not subject an AMR, the staff requested applicant to provide justification for the exclusion.

In its response, by letter dated September 20, 2006, the applicant stated:

> As described in LRA Section 2.3.3.9, the cable vault and switchgear rooms are protected by fully automatic total flooding CO_2 suppression systems initiated by ionization detectors. Bottles located in the west switchgear room may also provide a backup or second shot to the cable vault if desired. The diesel fire pump FO storage tank room is protected by a total flooding CO_2 suppression system initiated by heat detectors.
>
> As further described in LRA Section 2.3.3.9, the fire protection–CO_2 system is within the scope of license renewal and has the following intended function as required by 10 CFR 54.4(a)(3).
>
> - Provide the capability to extinguish fires in vital areas of the plant (10 CFR 50.48).
>
> The cable vault is the area referred to in the SE as the cable spreading area and battery room. Therefore, the CO_2 systems for the cable spreading area, battery room, and diesel driven fire water pump tank room are in-scope and subject to an AMR.

Based on its review, the staff found the applicant's response to RAI 2.3.3.9-1 acceptable because the total flooding CO_2 systems for the cable spreading area, battery room, and diesel driven fire water pump tank room were identified to be in the scope of license renewal and subject to an AMR. Further, the applicant clarified that the cable vault is the area referred to in the VYNPS fire protection safety evaluation as the cable spreading area and battery room. Therefore, the staff concludes that the total flooding CO_2 systems for the cable spreading area, battery room, and diesel driven fire water pump tank room and the associated components are correctly included in the scope of license renewal and subject to an AMR. The staff's concern described in RAI 2.3.3.9-1 is resolved.

In RAI 2.3.3.9-2, dated August 15, 2006, the staff stated that LRA Table 2.3.3-9 excludes several types of CO_2 fire suppression system components that appear in the VYNPS fire protection safety evaluation and its supplements and/or UFSAR, and which also appear in the LRA drawings colored in purple. These components are listed below.

- strainer housings
- pipe fittings
- pipe supports
- couplings
- odorizer
- threaded connections
- flexible hose
- latch door pull box
- pneumatic actuators
- CO_2 bottles (CO_2 storage cylinders)

For each, determine whether the component should be included in Table 2.3.3.9, and if not, the staff requested applicant justify the exclusion.

In its response, by letter dated September 20, 2006, the applicant stated:

- strainer housings – The CO_2 fire protection storage tank (TK-115-1) recirculation heater pump suction strainer (S-76-3) shown on LRA drawing LRA-G-191163-SH-03-0 has both filtration and pressure boundary functions. The strainer and its housing are both included in the "Strainer" line item in LRA Table 2.3.3-9.

- pipe fittings – As stated in LRA Section 2.0, the term "piping" in component lists may include pipe, pipe fittings (such as elbows and reducers), flow elements, orifices, and thermowells. Pipe fittings are included in the "Piping" line item in LRA Table 2.3.3-9.

- pipe supports – Since they support criterion (a)(3) equipment, piping supports are included in the structural AMR. They are included in the "Component and piping supports" line item in LRA Table 2.4-6.

- couplings – As stated in LRA Section 2.0, the term "piping" in component lists may include pipe, pipe fittings (such as elbows and reducers), flow elements, orifices, and thermowells. Couplings are pipe fittings included in the "Piping" line item in LRA Table 2.3.3-9.

- odorizer – Odorizer cylinders (OC-700, 701, 702, and 703) on switchgear room discharge lines are shown on LRA drawing LRA-G-191163-SH-03-0. The odorizer cylinders are included in the "Tank" line item in LRA Table 2.3.3-9.

- threaded connections – As stated in LRA Section 2.0, the term "piping" in component lists may include pipe, pipe fittings (such as elbows and reducers), flow elements, orifices, and thermowells. Threaded connections are pipe fittings included in the "Piping" line item in LRA Table 2.3.3-9.

- flexible hose – Hoses are replaced on a specified schedule and therefore, are not subject to an AMR as required by 10 CFR 54.21(a)(1)(ii).

- latch door pull box – This response assumes the reviewer means emergency manual release stations to initiate CO_2 flow. Manual release stations are active components; not subject to an AMR.

- pneumatic actuators – Pneumatic actuators (discharge delay timers) on deluge valves for the switchgear rooms are shown on LRA drawing LRA-G-191163-SH-03-0. Since the actuator subcomponents have a pressure boundary function, they are included in the line items for "Tank," "Valve body," and "Tubing" in Table 2.3.3-9.

- CO_2 bottles (CO_2 storage cylinders) – The CO_2 bottles, or storage cylinders, are included in the line item "Tank" in Table 2.3.3-9.

2-94

Based on its review, the staff found the applicant's response to RAI 2.3.3.9-2 acceptable. Although the applicant states that they consider these components to be included in other line items, the LRA descriptions of the line items do not specifically list all these components. The applicant identified the following components to be included in other line items in the scope of license renewal and subject to an AMR: strainer housings, pipe fittings, pipe supports, couplings, odorizer, threaded connections, pneumatic actuators, and CO_2 bottles. The staff is assured that the listed components will be considered appropriately during plant aging management activities. The staff found that the following components were not included in the line item descriptions in the LRA: flexible hoses and latch door pull box (emergency manual release stations to initiate CO_2 flow). The staff recognizes the applicant's interpretation of these components as active or short-lived components, which will result in more vigorous oversight of the condition and performance of the components. Because the applicant has interpreted these components are active, the staff concludes that the components were correctly excluded from the scope of license renewal and are not subject to an AMR. Therefore, the staff's concern described in RAI 2.3.3.9-2 is resolved.

In RAI 2.3.3.9-3, dated August 15, 2006, the staff stated that LRA Table 2.3.3-9 listed nozzles with an intended function of flow control as in-scope and subject to an AMR. Nozzles with intended functions of total flood, vent, and S nozzles are not listed. The staff requested that the applicant explain why these nozzles are not subject to an AMR.

In its response, by letter dated September 20, 2006, the applicant stated:

> The total flood nozzles in the CO_2 system are subject to an AMR, as indicated on drawings LRA-G-191163-SH-03-0 and LRA-G-191163-SH-04-0. They are included in the "Nozzle" line item in Table 2.3.3-9. As shown on the LRA drawings the CO_2 system does not have vent or S nozzles.

Based on its review, the staff finds the applicant's response to RAI 2.3.3.9-3 acceptable because it adequately explains that the flood nozzles in question are within the scope of license renewal and subject to an AMR. Further, the applicant stated that the flood nozzles are represented in the LRA Table 2.3.3-9 by the component type "Nozzles," and the CO_2 system does not have vent or S nozzles. Therefore, the staff's concern described in RAI 2.3.3.9-3 is resolved.

2.3.3.9.3 Conclusion

The staff reviewed the LRA to determine whether the applicant failed to identify any SSCs within the scope of License renewal or subject to an AMR. The staff finds no such omissions. On the basis of its review, the staff concludes that there is reasonable assurance that the applicant has adequately identified the fire protection-CO_2 system components within the scope of license renewal, as required by 10 CFR 54.4(a), and those subject to an AMR, as required by 10 CFR 54.21(a)(1).

2.3.3.10 Heating, Ventilation, and Air Conditioning

2.3.3.10.1 Summary of Technical Information in the Application

LRA Section 2.3.3.10 describes the heating, ventilation, and air conditioning (HVAC) and the house HB systems. The purpose of the HVAC system is to maintain the general area environment for personnel and equipment. It consists of several ventilation systems serving ten different areas of the plant: (1) primary containment ventilation normally operates to maintain drywell ambient temperature within acceptable ranges, (2) reactor building ventilation provides filtration and controls temperature, humidity, and migration of air from clean areas to areas of higher contamination, including exhaust to the plant stack. It also purges the drywell, (3) turbine building ventilation provides filtration and controls temperature, humidity, and migration of air from clean areas to areas of higher contamination. It exhausts building air to the plant stack (normal intake and exhaust function) in a monitored release path, (4) DG room ventilation supports operation of the EDGs, (5) control building ventilation maintains the environment in the main control room, (6) service building ventilation provides filtration, controls temperature and humidity, and exhausts potential contaminants to the plant stack. It maintains the hydrogen concentration well below 2 percent by volume in the HVAC equipment room (hydrogen is potentially generated from the AS-1 batteries), (7) radwaste building ventilation provides filtration (including filtration of exhaust sent to the plant stack) and controls temperature, humidity, and migration of air from clean areas to areas of higher contamination, (8) augmented off-gas building ventilation provides filtration (including filtration of exhaust sent to the plant stack) and temperature and humidity control, (9) intake structure ventilation maintains an environment suitable for operating personnel and equipment, including the diesel-driven fire pump, and (10) JDD building ventilation cools the JDD, which provides emergency lighting credited in the Appendix R safe shutdown capability assessment. The purpose of the HB system is to provide a source of steam for space heating and process requirements during all phases of station operation and heats the control room during normal operation. The system has two 50-percent boilers, various heaters, steam traps, valves, and piping.

The HVAC and HB systems have safety-related components relied upon to remain functional during and following DBEs. The failure of nonsafety-related systems SSCs potentially could prevent the satisfactory accomplishment of a safety-related function. In addition, the systems perform functions that support fire protection.

LRA Tables 2.3.3-10, 2.3.3-13-18, and 2.3.3-13-21 identify the following HVAC and HB system component types within the scope of license renewal and subject to an AMR:

- bolting
- compressor housing
- damper housing
- duct
- duct flexible connection
- expansion joint
- fan housing
- filter housing
- heat exchanger (fins)
- heat exchanger (housing)

- heat exchanger (shell)
- heater housing
- humidifier housing
- louver housing
- piping
- pump casing
- sight glass
- steam trap
- strainer
- strainer housing
- tank
- tubing
- valve body

The HVAC and HB system component intended functions within the scope of license renewal include the following:

- filtration
- heat transfer
- pressure boundary

2.3.3.10.2 Staff Evaluation

The staff reviewed LRA Section 2.3.3.10 and UFSAR Sections 5.2.3.7, 5.3.5, 10.7.6, and 10.12 using the evaluation methodology described in SER Section 2.3 and the guidance in SRP-LR Section 2.3.

The staff evaluated the system functions described in the LRA and UFSAR to verify that the applicant has not omitted from the scope of license renewal any components with intended functions as required by 10 CFR 54.4(a). The staff then reviewed those components that the applicant has identified as within the scope of license renewal to verify that the applicant has not omitted any passive and long-lived components subject to an AMR as required by 10 CFR 54.21(a)(1).

2.3.3.10.3 Conclusion

The staff reviewed the LRA to determine whether the applicant failed to identify any SSCs within the scope of license renewal or subject to an AMR. The staff finds no such omissions. On the basis of its review, the staff concludes that there is reasonable assurance that the applicant has adequately identified the HVAC and HB system components within the scope of license renewal, as required by 10 CFR 54.4(a), and those subject to an AMR, as required by 10 CFR 54.21(a)(1).

2.3.3.11 Primary Containment Atmosphere Control / Containment Atmosphere Dilution

2.3.3.11.1 Summary of Technical Information in the Application

LRA Section 2.3.3.11 describes the PCAC system, the containment atmosphere dilution (CAD) system, and the post-accident sampling system (PASS). The purpose of the PCAC system is to ensure that the containment atmosphere is inerted with N_2 during station power operation. The PCAC system establishes and maintains the required differential pressure between the drywell and torus. System instrumentation monitors key drywell and torus parameters, including temperature, pressure, moisture, drywell to torus differential pressure, and torus water level. The CAD system limits the concentration of oxygen in the primary containment so ignition of hydrogen and oxygen from a metal-water reaction following a LOCA will not occur. The PASS is included in this evaluation. The purpose of PASS is to provide representative samples of reactor coolant indicative of the extent and development of core damage.

The PCAC system, CAD system, and PASS have safety-related components relied upon to remain functional during and following DBEs. The failure of nonsafety-related SSCs in the system potentially could prevent the satisfactory accomplishment of a safety-related function.

LRA Tables 2.3.3-11, 2.3.3-13-3, 2.3.3-13-27, and 2.3.3-13-28 identify the following PCAC system, CAD system, and PASS component types within the scope of license renewal and subject to an AMR:

- bolting
- diaphragm
- dryer
- duct
- filter housing
- heat exchanger
- orifice
- piping
- pump casing
- tank
- trap
- tubing
- valve body

The component intended functions within the scope of license renewal include the following:

- flow control
- heat transfer
- pressure boundary

2.3.3.11.2 Staff Evaluation

The staff reviewed LRA Section 2.3.3.11 and UFSAR Sections 5.2.3.6, 5.2.6, 5.2.7, and 10.20 using the evaluation methodology described in SER Section 2.3 and the guidance in SRP-LR Section 2.3.

In conducting its review, the staff evaluated the system functions described in the LRA and UFSAR to verify that the applicant has not omitted from the scope of license renewal any components with intended functions as required by 10 CFR 54.4(a). The staff then reviewed those components that the applicant has identified as within the scope of license renewal to verify that the applicant has not omitted any passive and long-lived components subject to an AMR as required by 10 CFR 54.21(a)(1).

The staff's review of LRA Section 2.3.3.11 identified areas in which additional information was necessary to complete the review of the applicant's scoping and screening results. The applicant responded to the staff's RAIs as discussed below.

In RAI 2.3.3.11-1 dated August 16, 2006, the staff stated that license renewal drawing LRA-VY-E-75-002-0, at location K-13, penetration X209D to the H_2O_2 analyzers, shows a section of pipe to be within the scope of license renewal. However, this same section of pipe on drawing LRA-G-191165-0, at location E-16 from penetration X209D through the continuation to drawing LRA-VY-E-75-002-0, is not shown to be within the scope of license renewal. The staff requested that the applicant confirm that this section of pipe is within the scope of license renewal, or if not, justify its exclusion.

In its response dated September 20, 2006, the applicant stated that the section of pipe shown on license renewal drawing LRA-VY-E-75-002-0, at location K-13 at penetration X209D to the H_2O_2 analyzers and on drawing LRA-G-191165-0, at location E-16 from penetration X209D through the continuation to drawing LRA-VY-E-75-002-0 is within the scope of license renewal and subject to an AMR. Dashed lines (or phantom lines) on the drawings indicate that the actual line is shown on its primary system drawing. Phantom lines are not highlighted on the license renewal drawings.

Based on its review, the staff found the applicant's response to RAI 2.3.3.11-1 acceptable because the applicant confirmed that containment atmosphere dilution system piping 1"-VG-122-D1 connecting the H_2O_2 analyzers to the torus through penetration X-209D is within the scope of license renewal and subject to an AMR. Therefore, the staff concern described in RAI 2.3.3.11-1 is resolved.

In RAI 2.3.3.11-2 dated August 16, 2006, the staff stated that license renewal drawing LRA-VY-E-75-002-0, at location J-9 shows a pipe section, including valve NG-16 to pipe section 20-AC-13, within the scope of license renewal. However, this same section of pipe on drawing LRA-G-191175-SH-01-0, at location K-10 is not shown within the scope of license renewal. The staff requested that the applicant confirm that this section of pipe is within the scope of license renewal, or if not, to justify its exclusion.

In its response dated September 20, 2006, the applicant stated that the section of pipe shown on license renewal drawing LRA-VY-E-75-002-0, at location J-9, including valve NG-16 to pipe section 20-AC-13 and on drawing LRA-G-191175-SH-01-0, at location K-10 is within the scope of license renewal and subject to an AMR. Dashed lines (or phantom lines) on the drawings indicate that the actual line is shown on its primary system drawing. Phantom lines are not highlighted on the license renewal drawings.

Based on its review, the staff found the applicant response to RAI 2.3.3.11-2 acceptable because the applicant confirmed that containment atmosphere dilution system piping from primary containment and atmosphere control system piping 20"- AC-13 to valve NG-16 (1" NG-101A-EIN2) is within the scope of license renewal and subject to an AMR. Therefore, the staff concern described in RAI 2.3.3.11-2 is resolved.

In RAI 2.3.3.11-3 dated August 16, 2006, the staff stated that license renewal drawing LRA-VY-E-75-002-0, at location G-7 provides a continuation from valve VG-77 to drawing LRA-G-191165-0 (at location B-17) that is within the scope of license renewal. However, the license renewal boundary could not be located on drawing LRA-G-191165-0 (at location B-17). The staff requested that the applicant provide additional information for the continuation of this pipe section to the license renewal boundary and justify the boundary locations with respect to the applicable requirements of 10 CFR 54.4(a).

In its response dated September 20, 2006, the applicant stated that license renewal drawing LRA-VY-E-75-002-0, at location G-17 provides a continuation from valve VG-77 to drawing LRA-G-191165-0 that is within the scope of license renewal. The drawing references location B-17 on drawing LRA-G-191165-0. The hydrogen/oxygen analyzers are shown at location H-14 on drawing LRA-G-191165-0. Therefore, the appropriate reference location for the continuation on drawing LRA-G-191165-0 is H-14. An engineering request was submitted to correct the discrepancy on license renewal drawing LRA-VY-E-75-002-0. The piping to VG-77 is connected to ¾" pipe VG-109-TI prior to valve VG-20. As shown on the drawings, all of the piping and components from the primary containment air space to the analyzers and from the analyzers to the torus are within the scope of license renewal and subject to an AMR.
Based on its review, the staff found the applicant response to RAI 2.3.3.11-3 acceptable because the applicant provided appropriate documentation to demonstrate that piping upstream of valve VG-77 was connected to primary containment sample system line 3/4" VG-109-T1, piping and components were correctly identified within the scope of license renewal, and license renewal boundaries were appropriately identified on the sampling system flow diagram, LRA-G-191165-0. Therefore, the staff concern described in RAI 2.3.3.11-3 is resolved.

In RAI 2.3.3.11-4 dated August 16, 2006, the staff stated that license renewal drawing LRA-VY-E-75-002-0, at location J-18 shows a pipe section downstream of valve VG30A within the scope of license renewal. A drawing continuation to the license renewal boundary was not provided. The staff requested that the applicant provide additional information for the continuation of this pipe section to the license renewal boundary and justify the boundary locations with respect to the applicable requirements of 10 CFR 54.4(a).

In its response dated September 20, 2006, the applicant stated that license renewal drawing LRA-VY-E-75-002-0 shows hydrogen/oxygen analyzer panel SII within a dotted rectangular box at locations H-17 through J-18. Above the box, at location G-18, VG-29A is shown going to hydrogen/oxygen analyzer panel SI, which is not shown but is the same as the SII panel. Valve VG-30A, below the box at location J-18, is coming back from the SI panel. As shown on the drawing, all of the piping and components from the analyzer panels to the torus are within the scope of license renewal and subject to an AMR.

Based on its review, the staff found the applicant response to RAI 2.3.3.11-4 acceptable because the applicant adequately identified the piping and components in the H_2O_2 analyzer SAH-VG-5A SI panel which are within the scope of license renewal and subject to an AMR. These components were identified as those corresponding to components identified in panel SII on drawing LRA-VY-E-75-002-0. Therefore, the staff concern described in RAI 2.3.3.11-4 is resolved.

In RAI 2.3.3.11-5 dated August 16, 2006, the staff stated that license renewal drawing LRA-VY-191165-0, at location I-15 provides a continuation of a pipe section from the H_2O_2 analyzers to drawing LRA-VY-E-75-002-0 that is within the scope of license renewal. However, the license renewal boundary could not be located on drawing LRA-VY-E-75-002-0. The staff requested that the applicant provide additional information for the continuation of this pipe section to the license renewal boundary and justify the boundary locations with respect to the applicable requirements of 10 CFR 54.4(a).

In its response dated September 20, 2006, the applicant stated that an engineering request was submitted to correct the license renewal drawing discrepancies. Also, as shown on the drawings, all of the piping and components from the primary containment air space to the analyzers and from the analyzers to the torus are within the scope of license renewal and subject to an AMR.

Based on its review, the staff found the applicant response to RAI 2.3.3.11-5 acceptable because the applicant confirmed that sample system piping located on drawing LRA-G-191165-0, at location I-15 and H-14, is continued on drawing LRA-VY-E-75-002-0. Additionally, the applicant demonstrated these components and all of the piping and components from the primary containment air space to the analyzers and from the analyzers to the torus are within the scope of license renewal and subject to an AMR. Therefore, the staff concern described in RAI 2.3.3.11-5 is resolved.

In RAI 2.3.3.11-6 dated August 16, 2006, the staff stated that license renewal drawing LRA-VY-191165-0, at location C-12 provides continuations to drawing LRA-G-191267 (at locations H-12 and H-5) for two pipe lines from the post-accident sampling panel that are within the scope of license renewal. The license renewal boundary could not be located on LRA-G-191267-SH-01-0. The staff requested that the applicant provide additional information for the continuation of these pipe sections to the license renewal boundary and justify the boundary locations with respect to the applicable requirements of 10 CFR 54.4(a).

In its response dated September 20, 2006, the applicant confirmed that the two pipe lines from the post-accident sampling panel shown on license renewal drawing LRA-VY-191165-0, at location C-12 are continued on drawing LRA-G-191267-SH-01-0 (at location H-12 and H-5). The lines are depicted as "TYPICAL FOR FT63A" and "TYPICAL FOR FT63C" with reference to FT63B and FT63D piping which are identified within dashed rectangles on drawing LRA-G-191267-SH-01-0 at the specified locations. The table on drawing LRA-G-191267-SH-02-0, at location A-16, notes the instrument root valves associated with each jet pump. Drawing LRA-G-191267-SH-01-0 identifies the piping and components from the jet pump to the instruments as being within the scope of license renewal and subject to an AMR as part of the RCS pressure boundary described in LRA Section 2.3.1.3. Drawing LRA-G-191165-0 shows piping continuing from jet pump instrument root valve V-20B (typical) to PASS valve 102 and 101 and from root valve V-20D (typical) to PASS valve 104 and 103. The applicant confirmed that components in the sample line are within the scope of license renewal and subject to an AMR as part of the post-accident sampling system as described in LRA Section 2.3.3.11. Therefore, in accordance with 10 CFR 54.4(a)(1), the entire reactor coolant pressure boundary out to the second isolation valve on the PASS sample lines is within the scope of license renewal and subject to an AMR.

Based on its review, the staff found the applicant response to RAI 2.3.3.11-6 acceptable because the applicant submitted appropriate documentation acknowledging that all piping and components associated with primary containment atmosphere control and containment atmosphere dilution are within the scope of license renewal and subject to an AMR including all the reactor coolant pressure boundary up to and including the second post-accident sampling system (PASS) isolation valves. Therefore, the staff concern described in RAI 2.3.3.11-6 is resolved.

2.3.3.11.3 Conclusion

The staff reviewed the LRA accompanying license renewal drawings, and RAI responses to determine whether the applicant failed to identify any SSCs within the scope of license renewal or subject to an AMR. The staff finds no such omissions. On the basis of its review, the staff concludes that there is reasonable assurance that the applicant has adequately identified the PCAC system, CAD system, and PASS components within the scope of license renewal, as required by 10 CFR 54.4(a), and those subject to an AMR, as required by 10 CFR 54.21(a)(1).

2.3.3.12 John Deere Diesel

2.3.3.12.1 Summary of Technical Information in the Application

LRA Section 2.3.3.12 describes the JDD as a nonsafety-related skid-mounted engine powering a generator that supplies back-up electric power to plant lighting. It is located in a separate structure, the JDD building. The diesel is started electrically with batteries and does not require cooling water from other plant systems. Its license renewal purpose is to provide power to lighting panels credited as emergency lighting in the Appendix R safe shutdown capability analysis.

The JDD performs functions that support fire protection.

LRA Table 2.3.3-12 identifies the following JDD component types within the scope of license renewal and subject to an AMR:

- bolting
- expansion joint
- filter housing
- heat exchanger (radiator)
- heat exchanger (shell)
- heat exchanger (tubes)
- heater housing
- piping

- pump casing
- silencer
- tubing
- turbocharger

The JDD component intended functions within the scope of license renewal include the following:

- heat transfer
- pressure boundary

2.3.3.12.2 Staff Evaluation

The staff reviewed LRA Section 2.3.3.12 using the Tier-2 evaluation methodology described in SER Section 2.3 and the guidance in SRP-LR Section 2.3.

In conducting its review, the staff evaluated the system functions described in the LRA to verify that the applicant has not omitted from the scope of license renewal any components with intended functions as required by 10 CFR 54.4(a). The staff then reviewed those components that the applicant has identified as within the scope of license renewal to verify that the applicant has not omitted any passive and long-lived components subject to an AMR as required by 10 CFR 54.21(a)(1).

The staff's review of LRA Section 2.3.3.12 identified areas in which information provided in the LRA needed to be confirmed by the NRC Regional Inspection Team to complete the review of the applicant's scoping and screening results.

Inspection Item 2.3.3.12-1

LRA Section 2.3.3.12 indicts that the John Deere Diesel is installed in compliance with 10 CFR 50, Appendix R, requirements. However, due to a lack of available drawings and/or detailed description of the diesel equipment listed in LRA Table 2.3.3-12, it is difficult to determine if any AMR category components may have been omitted from the table. It is recommended that the JDD be inspected to assure all AMR category components are included in the list of LRA Table 2.3.3-12. The staff requested that the NRC Regional Inspection Team perform an inspection to ensure that the license renewal scope boundaries for these components satisfy the requirements of 10 CFR 54.4(a) (3). The staff identified this as confirmatory item 2.3.3.12-1.

In Inspection Report 05000271/2007006, Vermont Yankee Nuclear Power Station - NRC License Renewal Inspection Report, dated June 4, 2007, Attachment, Review of Safety Evaluation Report Confirmatory Items, the NRC Regional Inspection Team stated that the John Deere diesel system components are listed in LRA Table 2.3.3-12 and the supporting fuel oil day tank, fiberglass underground storage tank, and supply lines are listed in LRA Table 2.3.3-6, "Fuel Oil System."

Based on its review, the staff found the above response acceptable because the NRC Regional Inspection Team verified that all components subject to an AMR are included in LRA Table 2.3.3-12 and LRA Table 2.3.3-6 and confirmed that no other portions of the John Deere diesel system should have been included within scope based on 10 CFR 54.4(a)(3). Therefore, the staff concern described in confirmatory item 2.3.3.12-1 is resolved.

2.3.3.12.3 Conclusion

The staff reviewed the LRA and confirmatory item response to determine whether the applicant failed to identify any SSCs within the scope of license renewal or subject to an AMR. The staff finds no such omissions. On the basis of its review, the staff concludes that there is reasonable assurance that the applicant has adequately identified the JDD components within the scope of license renewal, as required by 10 CFR 54.4(a), and those subject to an AMR, as required by 10 CFR 54.21(a)(1).

2.3.3.13 Miscellaneous Systems In-scope as required by 10 CFR 54.4(a)(2)

2.3.3.13.1 Summary of Technical Information in the Application

LRA Section 2.3.3.13 describes the miscellaneous systems within the scope of license renewal requirements of 10 CFR 54.4(a)(2). Such systems interact with safety-related systems in one of two ways: (1) a functional failure where the failure of a nonsafety-related SSC to perform its function impacts a safety function or (2) a physical failure where a safety function is impacted by the loss of structural or mechanical integrity of an SSC in physical proximity to a safety-related component.

LRA Section 2.3.3.13.1 states that functional failures of nonsafety-related SSCs which could impact a safety function were identified only for systems with components supporting the main condenser and MSIV leakage pathway. Two of these systems are the augmented off-gas (AOG) and sampling systems, which are not described elsewhere in the LRA. Descriptions of these systems follow.

2.3.3.13A Augmented Off-gas

2.3.3.13A.1 Summary of Technical Information in the Application

The AOG system collects, processes, and discharges radioactive gaseous wastes to the atmosphere through the plant stack during normal operation. The system reduces the released quantities of gaseous and particulate radioactive material from the site to levels as low as practical during normal operation. The AOG system has subsystems that dispose of gases from the main condenser air ejectors, the start-up vacuum pump, and the gland seal condenser. The various subsystems are monitored continuously for radiation.

The failure of nonsafety-related SSCs in the AOG system could prevent the satisfactory accomplishment of a safety-related function.

LRA Table 2.3.3.13-1 identifies the following AOG system component types within the scope of license renewal and subject to an AMR:

- bolting
- filter housing
- piping
- stream trap
- tank
- tubing
- valve body

The AOG system component intended function within the scope of license renewal is to provide a pressure boundary.

2.3.3.13A.2 Staff Evaluation

The staff reviewed LRA Section 2.3.3.13.1 and UFSAR Section 9.4 using the Tier-2 evaluation methodology described in SER Section 2.3. and the guidance described in SRP-LR Section 2.3.

In conducting its review, the staff evaluated the system functions described in the LRA and UFSAR to verify that the applicant has not omitted from the scope of license renewal any components with intended functions as required by 10 CFR 54.4(a). The staff then reviewed those components that the applicant has identified as within the scope of license renewal to verify that the applicant has not omitted any passive and long-lived components subject to an AMR as required by 10 CFR 54.21(a)(1).

The staff's review of LRA Section 2.3.3.13.1 identified areas in which information provided in the LRA needed to be confirmed by the NRC Regional Inspection Team to complete the review of the applicant's scoping and screening results.

Inspection Item 2.3.3.13a-1

The LRA states that the AOG system is within the scope of license renewal based on requirements of 10 CFR 54.4(a)(2) because of the potential for physical interaction with safety-related components described in LRA Table 2.3.3.13-A. The determination of whether a component meets the requirements of 10 CFR 54.4(a)(2) for physical interactions is based on where it is located in a building and its proximity to safety-related equipment or where a structural/seismic boundary exists. This information is not provided on license renewal drawings nor was a detailed description provided in the LRA. Consequently, any omission of AOG components subject to an AMR cannot be determined. The staff requested that the NRC Regional Inspection Team perform an inspection to ensure that the license renewal scope boundaries for these components meet the requirements of 10 CFR 54.4(a)(2) and all the components subject to an AMR are included in LRA Table 2.3.3-13-1. The staff identified this as confirmatory item 2.3.3.13a-1.

In Inspection Report 05000271/2007006, Vermont Yankee Nuclear Power Station - NRC License Renewal Inspection Report, dated June 4, 2007, Attachment, Review of Safety Evaluation Report Confirmatory Items, the NRC Regional Inspection Team noted LRA

Table 2.3.3.13-B states that the portion of the AOG system associated with the plant stack loop seal is subject to an AMR based on 10 CFR 54.4(a)(2) for physical interactions. Since the boundaries for the portion of the system as described in LRA Table 2.3.3.13-B were not well defined, in its letter dated July 30, 2007, the applicant amended the table to read "portion of the system inside the plant stack." The inspector walked down the remainder of the system and confirmed that no other portions of the system should have been included based on 10 CFR 54.4(a)(2).

Based on its review, the staff found the above response acceptable because the applicant amended LRA Table 2.3.3.13-B as appropriate and the NRC regional inspector walked down the remainder of the AOG system outside the plant stack and confirmed that no other portions of the system should have been included within scope based on 10 CFR 54.4(a)(2). Therefore, the staff concern described in confirmatory item 2.3.3.13a-1 is resolved.

2.3.3.13A.3 Conclusion

The staff reviewed the LRA, accompanying license renewal drawings, and confirmatory item response to determine whether the applicant failed to identify any SSCs within the scope of license renewal or subject to an AMR. The staff finds no such omissions. On the basis of its review, the staff concludes that there is reasonable assurance that the applicant has adequately identified the AOG system components within the scope of license renewal, as required by 10 CFR 54.4(a), and those subject to an AMR, as required by 10 CFR 54.21(a)(1).

2.3.3.13B Sampling

2.3.3.13B.1 Summary of Technical Information in the Application

The sampling system provides means for sampling and testing various process fluids in the station in centralized locations. Fluids and gases are sampled continuously or periodically from equipment or systems reflecting station performance.

The failure of nonsafety-related SSCs in the sampling system could prevent the satisfactory accomplishment of a safety-related function.

LRA Table 2.3.3.13-41 identifies the following sampling system component types within the scope of license renewal and subject to an AMR:

- bolting
- piping
- strainer housing
- tubing
- valve body

The sampling system component intended function within the scope of license renewal is to provide a pressure boundary.

2.3.3.13B.2 Staff Evaluation

The staff reviewed LRA Section 2.3.3.13.1 and UFSAR Section 10.17 using the Tier-2 evaluation methodology described in SER Section 2.3. and the guidance described in SRP-LR Section 2.3.

The staff evaluated the system functions described in the LRA and UFSAR to verify that the applicant has not omitted from the scope of license renewal any components with intended functions as required by 10 CFR 54.4(a). The staff then reviewed those components that the applicant has identified as within the scope of license renewal to verify that the applicant has not omitted any passive and long-lived components subject to an AMR as required by 10 CFR 54.21(a)(1).

2.3.3.13B.3 Conclusion

The staff reviewed the LRA and accompanying license renewal drawings to determine whether the applicant failed to identify any SSCs within the scope of license renewal or subject to an AMR. The staff finds no such omissions. On the basis of its review, the staff concludes that there is reasonable assurance that the applicant has adequately identified the sampling system components within the scope of license renewal, as required by 10 CFR 54.4(a), and those subject to an AMR, as required by 10 CFR 54.21(a)(1).

Besides the augmented off-gas and sampling systems, other systems with components supporting the main condenser and MSIV leakage pathway where functional failures of nonsafety-related SSCs could impact a safety function are addressed in LRA Section 2.3.4.

LRA Table 2.3.3.13-A shows systems within the scope of license renewal with potential for physical interactions with safety-related components based on the criterion of 10 CFR 54.4(a)(2). Of these systems, the applicant stated that the following are not described elsewhere in the LRA:

- circulating water
- condensate demineralizer
- demineralized water
- equipment retired in place
- feedwater
- MG lube oil
- neutron monitoring
- potable water
- radwaste, liquid and solid
- reactor water clean-up
- RWCU filter demineralizer
- stator cooling

A description of each system above follows.

2.3.3.13C Condensate Demineralizer

2.3.3.13C.1 Summary of Technical Information in the Application

The condensate demineralizer (CD) system maintains the required purity of feedwater supplied to the reactor. The system minimizes corrosion product in the nuclear system so it does not affect fuel performance, nuclear system component accessibility, or the capacity required of the RWCU system. The CD system protects the nuclear system against the entry of foreign material due to condenser leaks. The system uses finely ground, mixed ion-exchange resins deposited upon the tubular elements of pressure precoat type filters (the filter-demineralizer units). The CD consist of five filter-demineralizer units (including an installed spare) operating in parallel. All are normally operated but sized so four units can support operation.

The failure of nonsafety-related SSCs in the CD system potentially could prevent the satisfactory accomplishment of a safety-related function.

LRA Table 2.3.3-13-4 identifies the following CD system component types within the scope of license renewal and subject to an AMR:

- bolting
- filter housing
- piping
- pump casing
- strainer housing
- tank
- tubing
- valve body

The CD system component intended function within the scope of license renewal is to provide pressure boundary.

2.3.3.13C.2 Staff Evaluation

The staff reviewed LRA Section 2.3.3.13.2 and UFSAR Section 11.7 using the evaluation methodology described in SER Section 2.3. and the guidance described in SRP-LR Section 2.3.

In conducting its review, the staff evaluated the system functions described in the LRA and UFSAR to verify that the applicant has not omitted from the scope of license renewal any components with intended functions as required by 10 CFR 54.4(a). The staff then reviewed those components that the applicant has identified as within the scope of license renewal to verify that the applicant has not omitted any passive and long-lived components subject to an AMR as required by 10 CFR 54.21(a)(1).

2.3.3.13C.3 Conclusion

The staff reviewed the LRA to determine whether the applicant failed to identify any SSCs within the scope of license renewal or subject to an AMR. The staff finds no such omissions. On the basis of its review, the staff concludes that there is reasonable assurance that the applicant has adequately identified the CD system components within the scope of license renewal, as required by 10 CFR 54.4(a), and those subject to an AMR, as required by 10 CFR 54.21(a)(1).

2.3.3.13D RWCU Filter Demineralizer

2.3.3.13D.1 Summary of Technical Information in the Application

The RWCU filter demineralizer (CUFD) system filters and cleans reactor water. The CUFD is the filter-demineralizer portion of the RWCU system and consists of the filter/demineralizer tanks, piping, and valves.

The failure of nonsafety-related SSCs in the CUFD system potentially could prevent the satisfactory accomplishment of a safety-related function.

LRA Table 2.3.3-13-8 identifies the following CUFD system component types within the scope of license renewal and subject to an AMR:

- bolting
- filter housing
- orifice
- piping
- pump casing
- sight glass
- strainer housing
- tank
- tubing
- valve body

The CUFD system component intended function within the scope of license renewal is to provide pressure boundary.

2.3.3.13D.2 Staff Evaluation

The staff reviewed LRA Section 2.3.3.13.2 and UFSAR Section 4.9 using the Tier-1 evaluation methodology described in SER Section 2.3. and the guidance described in SRP-LR Section 2.3.

In conducting its review, the staff evaluated the system functions described in the LRA and UFSAR to verify that the applicant has not omitted from the scope of license renewal any components with intended functions as required by 10 CFR 54.4(a). The staff then reviewed those components that the applicant has identified as within the scope of license renewal to verify that the applicant has not omitted any passive and long-lived components subject to an AMR as required by 10 CFR 54.21(a)(1).

2-109

2.3.3.13D.3 Conclusion

The staff reviewed the LRA to determine whether the applicant failed to identify any SSCs within the scope of license renewal or subject to an AMR. The staff finds no such omissions. On the basis of its review, the staff concludes that there is reasonable assurance that the applicant has adequately identified the CUFD system components within the scope of license renewal, as required by 10 CFR 54.4(a), and those subject to an AMR, as required by 10 CFR 54.21(a)(1).

2.3.3.13E Circulating Water

2.3.3.13E.1 Summary of Technical Information in the Application

The circulating water (CW) system is a heat sink for steam condensation for the main condensers. Heat removal in the condensers is accomplished by a continuous supply of cooling water pumped from and returned to the Connecticut River or by recirculation flow pumped through cooling towers by three vertical circulating water pumps in the intake structure. Trash racks and traveling water screens protect the circulating water pumps from debris. During cold weather, recirculation of water from the discharge structure to the intake structure prevents icing at the screens and intakes. Two cooling towers have the capacity to remove the total heat load from the circulating water. Three vertical circulating water booster pumps provide the necessary head for cooling tower operation and the recirculation mode.

The CW system has safety-related components relied upon to remain functional during and following DBEs. The failure of nonsafety-related SSCs in the CW system potentially could prevent the satisfactory accomplishment of a safety-related function.

LRA Table 2.3.3-13-9 identifies the following CW system component types within the scope of license renewal and subject to an AMR:

- bolting
- expansion joint
- piping
- pump casing
- tank
- tubing
- valve body

The CW system component intended function within the scope of license renewal is to provide pressure boundary.

2.3.3.13E.2 Staff Evaluation

The staff reviewed LRA Section 2.3.3.13.2, and UFSAR Sections 10.8, 11.6, and 11.9 using the Tier-2 evaluation methodology described in SER Section 2.3. and the guidance described in SRP-LR Section 2.3.

In conducting its review, the evaluated the system functions described in the LRA and UFSAR to verify that the applicant has not omitted from the scope of license renewal any components with intended functions as required by 10 CFR 54.4(a). The staff then reviewed those components that the applicant has identified as within the scope of license renewal to verify that the applicant has not omitted any passive and long-lived components subject to an AMR as required by 10 CFR 54.21(a)(1).

The staff's review of LRA Section 2.3.3.13 identified areas in which information provided in the LRA needed to be confirmed by the NRC Regional Inspection Team to complete the review of the applicant's scoping and screening results.

Inspection Item 2.3.3.13e-1

The LRA states that the circulating water system is within the scope of license renewal based on the potential for physical interaction with safety-related components as required by 10 CFR 54.4(a)(2) and described in LRA Table 2.3.3.13-A. The applicant did not provide drawings highlighting in-scope components required by 10 CFR 54.4(a)(2), stating that the drawings would not provide significant additional information because they do not indicate proximity of components to safety-related equipment and do not identify structural/seismic boundaries. Without license renewal drawings and/or detailed description of the circulating water system, the omission of components subject to an AMR cannot be determined (see LRA Table 2.3.3-13-9). The staff requested that the NRC Regional Inspection Team perform an inspection to ensure that the license renewal scope boundaries for these components satisfy the requirements of 10 CFR 54.4(a)(2) and all the components subject to an AMR are included in LRA Table 2.3.3-13-9. The staff identified this as confirmatory item 2.3.3.13e-1.

In Inspection Report 05000271/2007006, Vermont Yankee Nuclear Power Station - NRC License Renewal Inspection Report, dated June 4, 2007, Attachment, Review of Safety Evaluation Report Confirmatory Items, the NRC Regional Inspection Team stated that if any nonsafety-related portion of a fluid system is located within a building containing safety-related components, the components within the system are within the license renewal scope. Further, applicant's letter to the NRC dated July 3, 2007, LRA Amendment 27, Attachment 2 states that there are no nonsafety-related systems for which the applicant has not identified the nonsafety-related portions of systems which are attached to safety-related systems and required to be in the scope of license renewal in accordance with 10 CFR 54.4(a)(2). However, as a result of discussions with the staff during the Region I inspection (February 2007), the applicant determined that some safety-related SSCs in the VY turbine building required consideration for potential spatial impacts from nonsafety-related SSCs in accordance with 10 CFR 54.4(a)(2). Therefore, an expanded review for SSCs in the turbine building determined that additional components required an AMR. Those additional component types were added to LRA Table 2.3.3-13-9, as addressed in the applicant's letters to the staff dated July 30, 2007 and August 16, 2007.

Based on its review, the staff found the above response acceptable because the NRC Regional Inspection Team found that if any nonsafety-related portion of a fluid system is located within a building containing safety-related components, the components within the system are within the license renewal scope in accordance with 10 CFR 54.4(a)(2) but that there were spatial impact concerns from nonsafety-related SSCs in the turbine building. The additional component types have been added to LRA Table 2.3.3-13-9. Therefore, the staff concern regarding components of the CW system described in confirmatory item 2.3.3.13e-1 is resolved.

2.3.3.13E.3 Conclusion

The staff reviewed the LRA and the confirmatory item response to determine whether the applicant failed to identify any SSCs within the scope of license renewal or subject to an AMR. The staff finds no such omissions. On the basis of its review, the staff concludes that there is reasonable assurance that the applicant has adequately identified the CW system components within the scope of license renewal, as required by 10 CFR 54.4(a), and those subject to an AMR, as required by 10 CFR 54.21(a)(1).

2.3.3.13F Demineralized Water

2.3.3.13F.1 Summary of Technical Information in the Application

The demineralized water (DW) system provides treated makeup water for such plant components as the condensate storage tank, spent fuel pool, RBCCW, and turbine building closed cooling water systems. This supply function is not a safety function. The DW system consists of the demineralized water transfer system including the demineralized water storage tank, demineralized water transfer pumps, piping, and valves, but not including the condensate storage tank or CST system components.

The failure of nonsafety-related SSCs in the DW system potentially could prevent the satisfactory accomplishment of a safety-related function.

LRA Table 2.3.3-13-12 identifies the following DW system component types within the scope of license renewal and subject to an AMR:

- bolting
- orifice
- piping
- pump casing
- tank
- tubing
- valve body

The DW system component intended function within the scope of license renewal is to provide pressure boundary.

2.3.3.13F.2 Staff Evaluation

The staff reviewed LRA Section 2.3.3.13.2 and UFSAR Section 10.13.3 using the Tier-2 evaluation methodology described in SER Section 2.3. and the guidance described in SRP-LR Section 2.3.

The staff evaluated the system functions described in the LRA and UFSAR to verify that the applicant has not omitted from the scope of license renewal any components with intended functions as required by 10 CFR 54.4(a). The staff then reviewed those components that the applicant has identified as within the scope of license renewal to verify that the applicant has not omitted any passive and long-lived components subject to an AMR as required by 10 CFR 54.21(a)(1).

2.3.3.13F.3 Conclusion

The staff reviewed the LRA to determine whether the applicant failed to identify any SSCs within the scope of license renewal or subject to an AMR. The staff finds no such omissions. On the basis of its review, the staff concludes that there is reasonable assurance that the applicant has adequately identified the DW system components within the scope of license renewal, as required by 10 CFR 54.4(a), and those subject to an AMR, as required by 10 CFR 54.21(a)(1).

2.3.3.13G Feedwater

2.3.3.13G.1 Summary of Technical Information in the Application

The feedwater (FW) system provides demineralized water from the condensate system to the reactor vessel at a rate sufficient to maintain adequate reactor vessel water level. The FW system consists of three reactor feedwater pumps, four high-pressure feedwater heaters (two per train), valves, and piping.

The failure of nonsafety-related SSCs in the FW system potentially could prevent the satisfactory accomplishment of a safety-related function.

LRA Table 2.3.3-13-13 identifies the following FW system component types within the scope of license renewal and subject to an AMR:

- bolting
- heat exchanger (shell)
- orifice
- piping
- pump casing
- strainer housing
- tubing
- valve body

The FW system component intended function within the scope of license renewal is to provide pressure boundary.

2.3.3.13G.2 Staff Evaluation

The staff reviewed LRA Section 2.3.3.13.2 and UFSAR Section 11.8 using the Tier-2 evaluation methodology described in SER Section 2.3. and the guidance described in SRP-LR Section 2.3.

The staff evaluated the system functions described in the LRA and UFSAR to verify that the applicant has not omitted from the scope of license renewal any components with intended functions as required by 10 CFR 54.4(a). The staff then reviewed those components that the applicant has identified as within the scope of license renewal to verify that the applicant has not omitted any passive and long-lived components subject to an AMR as required by 10 CFR 54.21(a)(1).

2.3.3.13G.3 Conclusion

The staff reviewed the LRA to determine whether the applicant failed to identify any SSCs within the scope of license renewal or subject to an AMR. The staff finds no such omissions. On the basis of its review, the staff concludes that there is reasonable assurance that the applicant has adequately identified the FW system components within the scope of license renewal, as required by 10 CFR 54.4(a), and those subject to an AMR, as required by 10 CFR 54.21(a)(1).

2.3.3.13H MG Lube Oil

2.3.3.13H.1 Summary of Technical Information in the Application

The MGLO system lubricates the reactor recirculation pump motor generator set during its operation. The MGLO system has lube oil pumps, heat exchangers, piping, and valves.

The failure of nonsafety-related SSCs in the MGLO system potentially could prevent the satisfactory accomplishment of a safety-related function.

LRA Table 2.3.3-13-23 identifies the following MGLO system component types within the scope of license renewal and subject to an AMR:

- bolting
- heat exchanger (shell)
- piping
- pump casing
- tubing
- valve body

The MGLO system component intended function within the scope of license renewal is to provide pressure boundary.

2.3.3.13H.2 Staff Evaluation

The staff reviewed LRA Section 2.3.3.13.2 and UFSAR Section 7.9.4.4.1 using the evaluation methodology described in SER Section 2.3. and guidance described in SRP-LR Section 2.3.

In conducting its review, the evaluated the system functions described in the LRA and UFSAR to verify that the applicant has not omitted from the scope of license renewal any components with intended functions as required by 10 CFR 54.4(a). The staff then reviewed those components that the applicant has identified as within the scope of license renewal to verify that the applicant has not omitted any passive and long-lived components subject to an AMR as required by 10 CFR 54.21(a)(1).

2.3.3.13H.3 Conclusion

The staff reviewed the LRA to determine whether the applicant failed to identify any SSCs within the scope of license renewal or subject to an AMR. The staff finds no such omissions. On the basis of its review, the staff concludes that there is reasonable assurance that the applicant has adequately identified the MGLO system components within the scope of license renewal, as required by 10 CFR 54.4(a), and those subject to an AMR, as required by 10 CFR 54.21(a)(1).

2.3.3.13I Neutron Monitoring

2.3.3.13I.1 Summary of Technical Information in the Application

The neutron monitoring (NM) system indicates neutron flux, which can be correlated to thermal power level, for the entire range of flux conditions in the core. The system consists of incore neutron detectors and out-of-core electronic monitoring equipment. The source-range and intermediate-range monitors indicate flux levels during reactor startup and lower power operation. The local-power range and average-power range monitors assess local and overall flux conditions during power range operation. Rod block monitors prevent rod withdrawal when reactor power should not be increased at the current reactor coolant flow rate. The traversing incore probe system calibrates individual neutron monitoring sensors. The safety function of the NM system is to detect conditions in the core that threaten the overall integrity of the fuel barrier by excessive power generation and to provide signals to the reactor protection system to limit the release of radioactive material from the fuel barrier.

The NM system has safety-related components relied upon to remain functional during and following DBEs. The failure of nonsafety-related SSCs in the NM system potentially could prevent the satisfactory accomplishment of a safety-related function.

LRA Table 2.3.3-13-26 identifies the following NM system component types within the scope of license renewal and subject to an AMR:

- piping
- tubing
- valve body

The NM system component intended function within the scope of license renewal is to provide pressure boundary.

2.3.3.13I.2 Staff Evaluation

The staff reviewed LRA Section 2.3.3.13.2, and UFSAR Sections 1.6.2.2, 1.6.4.1.3, and 7.5 using the evaluation methodology described in SER Section 2.3. The staff conducted its review in accordance with the guidance described in SRP-LR Section 2.3.

The staff evaluated the system functions described in the LRA and UFSAR to verify that the applicant has not omitted from the scope of license renewal any components with intended functions as required by 10 CFR 54.4(a). The staff then reviewed those components that the applicant has identified as within the scope of license renewal to verify that the applicant has not omitted any passive and long-lived components subject to an AMR as required by 10 CFR 54.21(a)(1).

2.3.3.13I.3 Conclusion

The staff reviewed the LRA to determine whether the applicant failed to identify any SSCs within the scope of license renewal or subject to an AMR. The staff finds no such omissions. On the basis of its review, the staff concludes that there is reasonable assurance that the applicant has adequately identified the NM system components within the scope of license renewal, as required by 10 CFR 54.4(a), and those subject to an AMR, as required by 10 CFR 54.21(a)(1).

2.3.3.13J Potable Water

2.3.3.13J.1 Summary of Technical Information in the Application

The potable water (PW) system supplies treated water suitable for drinking and for sanitary purposes to lavatories, service sinks, combination emergency showers and eyewashes, kitchen sinks, bench sinks, showers, and wall hydrants.

The failure of nonsafety-related SSCs in the PW system potentially could prevent the satisfactory accomplishment of a safety-related function.

LRA Table 2.3.3-13-29 identifies the following PW system component types within the scope of license renewal and subject to an AMR:

- bolting
- filter housing
- piping
- strainer housing
- tank
- valve body

The PW system component intended function within the scope of license renewal is to provide pressure boundary.

2.3.3.13J.2 Staff Evaluation

The staff reviewed LRA Section 2.3.3.13.2 and UFSAR Section 10.15 using the Tier-1 evaluation methodology described in SER Section 2.3. and the guidance described in SRP-LR Section 2.3.

In conducting its review, the staff evaluated the system functions described in the LRA and UFSAR to verify that the applicant has not omitted from the scope of license renewal any components with intended functions as required by 10 CFR 54.4(a). The staff then reviewed those components that the applicant has identified as within the scope of license renewal to verify that the applicant has not omitted any passive and long-lived components subject to an AMR as required by 10 CFR 54.21(a)(1).

2.3.3.13J.3 Conclusion

The staff reviewed the LRA to determine whether the applicant failed to identify any SSCs within the scope of license renewal or subject to an AMR. The staff finds no such omissions. On the basis of its review, the staff concludes that there is reasonable assurance that the applicant has adequately identified the PW system components within the scope of license renewal, as required by 10 CFR 54.4(a), and those subject to an AMR, as required by 10 CFR 54.21(a)(1).

2.3.3.13K Radwaste, Liquid and Solid

2.3.3.13K.1 Summary of Technical Information in the Application

The purpose of the liquid radwaste (RDW) system is to collect potentially radioactive liquid wastes, treats them, and returns the processed radioactive liquid wastes to the station for reuse. The solid RDW system collects and processes radioactive solid wastes for temporary onsite storage and offsite shipment for permanent disposal. The RDW system monitors the drywell floor and equipment drain sump pump-out rate for reactor coolant leak detection. The liquid portion of the RDW system consists of floor and equipment drains for handling tanks, piping, pumps, process equipment, instrumentation, and auxiliaries necessary to collect, process, store, and dispose of potentially radioactive wastes. A small portion of the system connected to the RHR system maintains the RHR system pressure boundary.

The RDW system has safety-related components relied upon to remain functional during and following DBEs. The failure of nonsafety-related SSCs in the RDW system potentially could prevent the satisfactory accomplishment of a safety-related function.

LRA Table 2.3.3-13-32 identifies the following RDW system component types within the scope of license renewal and subject to an AMR:

- bolting
- orifice
- piping
- pump casing
- strainer housing

- tank
- tubing
- valve body

The RDW system component intended function within the scope of license renewal is to provide pressure boundary.

2.3.3.13K.2 Staff Evaluation

The staff reviewed LRA Section 2.3.3.13.2, and UFSAR Sections 9.2 and 9.3 using the Tier-2 evaluation methodology described in SER Section 2.3. and the guidance described in SRP-LR Section 2.3.

In conducting its review, the staff evaluated the system functions described in the LRA and UFSAR to verify that the applicant has not omitted from the scope of license renewal any components with intended functions as required by 10 CFR 54.4(a). The staff then reviewed those components that the applicant has identified as within the scope of license renewal to verify that the applicant has not omitted any passive and long-lived components subject to an AMR as required by 10 CFR 54.21(a)(1).

2.3.3.13K.3 Conclusion

The staff reviewed the LRA to determine whether the applicant failed to identify any SSCs within the scope of license renewal or subject to an AMR. The staff finds no such omissions. On the basis of its review, the staff concludes that there is reasonable assurance that the applicant has adequately identified the RDW system components within the scope of license renewal, as required by 10 CFR 54.4(a), and those subject to an AMR, as required by 10 CFR 54.21(a)(1).

2.3.3.13L Equipment Retired in Place

2.3.3.13L.1 Summary of Technical Information in the Application

This system designation in the component database is for obsolete equipment. It has no safety-related components and no system intended functions; however, certain components supporting safety-related components are required to maintain structural integrity.

The failure of nonsafety-related SSCs of equipment retired in place (RIP) potentially could prevent the satisfactory accomplishment of a safety-related function.

LRA Table 2.3.3-13-35 identifies the following component types of equipment RIP within the scope of license renewal and subject to an AMR:

- bolting
- piping
- valve body

The equipment RIP component intended function within the scope of license renewal is to provide pressure boundary.

2.3.3.13L.2 Staff Evaluation

The staff reviewed LRA Section 2.3.3.13.2 using the Tier-1 evaluation methodology described in SER Section 2.3. and the guidance described in SRP-LR Section 2.3.

In conducting its review, staff evaluated the system functions described in the LRA and UFSAR to verify that the applicant has not omitted from the scope of license renewal any components with intended functions as required by 10 CFR 54.4(a). The staff then reviewed those components that the applicant has identified as within the scope of license renewal to verify that the applicant has not omitted any passive and long-lived components subject to an AMR as required by 10 CFR 54.21(a)(1).

2.3.3.13L.3 Conclusion

The staff reviewed the LRA to determine whether the applicant failed to identify any SSCs within the scope of license renewal or subject to an AMR. The staff finds no such omissions. On the basis of its review, the staff concludes that there is reasonable assurance that the applicant has adequately identified the equipment RIP components within the scope of license renewal, as required by 10 CFR 54.4(a), and those subject to an AMR, as required by 10 CFR 54.21(a)(1).

2.3.3.13M Reactor Water Clean-Up

2.3.3.13M.1 Summary of Technical Information in the Application

The RWCU system maintains high reactor water purity to limit chemical and corrosive action and to remove corrosion products to limit impurities available to activate neutron flux. The RWCU system purifies the reactor coolant water by continuously removing a portion of the reactor recirculation flow from the suction side of a recirculation pump, sending the removed flow through filter-demineralizer units to undergo mechanical filtration and ion exchange processes, and returning the processed fluid back to the reactor via the feedwater line.

The RWCU system has safety-related components relied upon to remain functional during and following DBEs. The failure of nonsafety-related SSCs in the RWCU system potentially could prevent the satisfactory accomplishment of a safety-related function.

LRA Table 2.3.3-13-36 identifies the following RWCU system component types within the scope of license renewal and subject to an AMR:

- bolting
- filter housing
- heat exchanger (shell)
- orifice
- piping
- pump casing

- strainer housing
- tank
- tubing
- valve body

The RWCU system component intended function within the scope of license renewal is to provide pressure boundary.

2.3.3.13M.2 Staff Evaluation

The staff reviewed LRA Section 2.3.3.13.2 and UFSAR Section 4.9 using the Tier-2 evaluation methodology described in SER Section 2.3. and the guidance described in SRP-LR Section 2.3.

In conducting its review, the staff evaluated the system functions described in the LRA and UFSAR to verify that the applicant has not omitted from the scope of license renewal any components with intended functions as required by 10 CFR 54.4(a). The staff then reviewed those components that the applicant has identified as within the scope of license renewal to verify that the applicant has not omitted any passive and long-lived components subject to an AMR as required by 10 CFR 54.21(a)(1).

The staff's review of LRA Section 2.3.3.13.2 identified an area in which additional information was necessary to complete the review of the applicant's scoping and screening results. The applicant responded to the staff's RAI as discussed below.

The staff stated that license renewal drawing LRA-G-191178-SH-01-0, at location D-4, shows the common elbow differential flow element upstream piping and high side instrument lines connected to flow transmitters FT-12-1A and FT-12-1 B as not within the scope of license renewal. A failure of the flow element upstream RWCU piping or common high side instrument line could prevent the flow transmitters from detecting a high flow condition and the subsequent auto isolation of the RWCU isolation valves. The UFSAR states that the high flow auto closure of the RWCU isolation valves prevents excessive loss of reactor coolant and reduces the amount of radioactive material released from the nuclear system caused by an RWCU line break. In RAI 2.3.3.13k-1 dated August 16, 2006, the staff requested that the applicant confirm whether the RWCU high flow auto isolation will occur when negative differential pressure is caused by either failure of the flow element upstream piping or the common high side instrument line. If not, explain why the flow element upstream piping and the common high side instrument lines are not shown to be within the scope of license renewal.

In its response dated September 20, 2006, the applicant stated that the flow element upstream piping and the common high side instrument lines are within the scope of license renewal based on the requirements of 10 CFR 54.4(a)(2) and thus are not shown as highlighted on the drawing. As stated in LRA Table 2.3.3.1B, "Description of Nonsafety-Related System Components Subject to Aging Management Review Based on 10 CFR 54.4(a)(2) for Physical Interactions," the nonsafety-related portion of the RWCU system located inside the reactor building is within the scope of license renewal and subject to an AMR. The common elbow differential flow element upstream piping and high side instrument lines connected to flow transmitters FT-12-1A and FT-12-1B are located inside the reactor building and are included in Table 2.3.3-13-36, "Reactor Water Clean-Up (RWCU) System Nonsafety-Related Systems and

Components Affecting Safety-Related Systems Components Subject to Aging Management Review." They are listed as component types of piping, tubing and valve body. As discussed in LRA Section 2.1.2.1.3, "Mechanical System Drawings," in-scope components required by 10 CFR 54.4(a)(2) are not highlighted on the drawings.

Based on its review, the staff found the applicant response to RAI 2.3.3.13k-1 acceptable because the applicant acknowledged that the flow element upstream piping and the common high side instrument lines connected to flow transmitters FT-12-1A and FT-12-1B are within the scope of license renewal and subject to an AMR based on the potential for physical interaction with safety-related systems in accordance with 10 CFR 54.4(a)(2). Therefore, the staff concern described in RAI 2.3.3.13k-1 is resolved.

The staff's review of LRA Section 2.3.3.13.2 identified areas in which information provided in the LRA needed to be confirmed by the NRC Regional Inspection Team to complete the review of the applicant's scoping and screening results.

Inspection Item 2.3.3.13m-1

The LRA states that the RWCU system is within the scope of license renewal in accordance with 10 CFR 54.4(a)(2) because of the potential for physical interaction with safety-related components as described in LRA Table 2.3.3.13-A. The determination of whether a component meets the requirements of 10 CFR 54.4(a)(2) for physical interactions is based on where it is located in a building and its proximity to safety-related equipment or where a structural/seismic boundary exists. This information is not provided on license renewal drawings nor was a detailed description provided in the LRA. Consequently, any omission of RWCU components subject to an AMR cannot be determined. The staff requested that the NRC Regional Inspection Team perform an inspection to ensure that the license renewal scope boundaries for these components satisfy the requirements of 10 CFR 54.4(a)(2) and all the components subject to an AMR are included in LRA Table 2.3.3-13-36. The staff identified this as confirmatory item 2.3.3.13m-1.

In Inspection Report 05000271/2007006, Vermont Yankee Nuclear Power Station - NRC License Renewal Inspection Report, dated June 4, 2007, Attachment, Review of Safety Evaluation Report Confirmatory Items, the NRC Regional Inspection Team stated that if any nonsafety-related portion of a fluid system is located within a building containing safety-related components, the components within the system are within the license renewal scope. Further, the applicant's letter to the NRC dated July 3, 2007, LRA Amendment 27, Attachment 2 states that there are no nonsafety-related systems for which the applicant has not identified the nonsafety-related portions of systems which are attached to safety-related systems and required to be in the scope of license renewal in accordance with 10 CFR 54.4(a)(2). The applicant also stated that there were no additional components that should be within scope based on 10 CFR 54.4(a) as identified during the NRC Regional Inspection and subsequent applicant reviews.

Based on its review, the staff found the above response acceptable because the NRC Regional Inspection Team found that if any nonsafety-related portion of a fluid system is located within a building containing safety-related components, the components within the system are within the license renewal scope in accordance with 10 CFR 54.4(a)(2) and that there were no additional components identified that should be in-scope based on 10 CFR 54.4(a). Therefore, the staff concern regarding the components of the RWCU system described in confirmatory item 2.3.3.13m-1 is resolved.

2.3.3.13M.3 Conclusion

The staff reviewed the LRA, RAI, and confirmatory item responses to determine whether the applicant failed to identify any SSCs within the scope of license renewal or subject to an AMR. The staff finds no such omissions. On the basis of its review, the staff concludes that there is reasonable assurance that the applicant has adequately identified the RWCU system components within the scope of license renewal, as required by 10 CFR 54.4(a), and those subject to an AMR, as required by 10 CFR 54.21(a)(1).

2.3.3.13N Stator Cooling

2.3.3.13N.1 Summary of Technical Information in the Application

The stator cooling system cools the stator winding of the main generator. The system permits generator load changes with minimum variation of stator winding temperature. The stator copper is in direct contact with low-conductivity water of automatically-controlled temperature and pressure; therefore, average copper temperature can be kept essentially constant, practically eliminating thermal stress cycling of the insulation.

The failure of nonsafety-related SSCs in the stator cooling system potentially could prevent the satisfactory accomplishment of a safety-related function.

LRA Table 2.3.3-13-39 identifies the following stator cooling system component types within the scope of license renewal and subject to an AMR:

- bolting
- cooler
- filter housing
- heat exchanger (shell)
- piping
- pump casing
- strainer housing
- tank
- tubing
- valve body

The stator cooling system component intended function within the scope of license renewal is to provide pressure boundary.

2.3.3.13N.2 Staff Evaluation

The staff reviewed LRA Section 2.3.3.13.2 and UFSAR Section 8.2.3.11.2 using the Tier-1 evaluation methodology described in SER Section 2.3. and the guidance described in SRP-LR Section 2.3.

In conducting its review, the staff evaluated the system functions described in the LRA and UFSAR to verify that the applicant has not omitted from the scope of license renewal any components with intended functions as required by 10 CFR 54.4(a). The staff then reviewed those components that the applicant has identified as within the scope of license renewal to verify that the applicant has not omitted any passive and long-lived components subject to an AMR as required by 10 CFR 54.21(a)(1).

2.3.3.13N.3 Conclusion

The staff reviewed the LRA to determine whether the applicant failed to identify any SSCs within the scope of license renewal or subject to an AMR. The staff finds no such omissions. On the basis of its review, the staff concludes that there is reasonable assurance that the applicant has adequately identified the stator cooling system components within the scope of license renewal, as required by 10 CFR 54.4(a), and those subject to an AMR, as required by 10 CFR 54.21(a)(1).

2.3.3.13O HD & HV Instruments

2.3.3.13O.1 Summary of Technical Information in the Application

The heater drain (HD) and the heater vent (HV) instruments system provides indication, alarm and control functions for associated systems (heater drains and heater vents).

The failure of nonsafety-related SSCs in the HD & HV instruments system potentially could prevent the satisfactory accomplishment of a safety-related function.

LRA Table 2.3.3-13-43 identifies the following HD & HV instruments system component types within the scope of license renewal and subject to an AMR:

*   bolting
*   piping
*   tubing
*   valve body

The HD & HV instruments system component intended function within the scope of license renewal is to provide pressure boundary.

2.3.3.13O.2 Staff Evaluation

The staff reviewed LRA Section 2.3.3.13.2 using the Tier-1 evaluation methodology described in SER Section 2.3 and the guidance described in SRP-LR Section 2.3.

In conducting its review, the staff evaluated the system functions described in the LRA to verify that the applicant has not omitted from the scope of license renewal any components with intended functions as required by 10 CFR 54.4(a). The staff then reviewed those components that the applicant has identified as within the scope of license renewal to verify that the applicant has not omitted any passive and long-lived components subject to an AMR as required by 10 CFR 54.21(a)(1).

2.3.3.13O.3 Conclusion

The staff reviewed the LRA to determine whether the applicant failed to identify any SSCs within the scope of license renewal or subject to an AMR. The staff finds no such omissions. On the basis of its review, the staff concludes that there is reasonable assurance that the applicant has adequately identified the HD & HV instruments system components within the scope of license renewal, as required by 10 CFR 54.4(a), and those subject to an AMR, as required by 10 CFR 54.21(a)(1).

2.3.3.13P Air Evacuation

2.3.3.13P.1 Summary of Technical Information in the Application

The air evacuation (AE) system evacuates gases from the main turbine and main condenser during startup and maintains them free of noncondensible gases during operation.

The failure of nonsafety-related SSCs in the AE system potentially could prevent the satisfactory accomplishment of a safety-related function.

LRA Table 2.3.3-13-44 identifies the following AE system component types within the scope of license renewal and subject to an AMR:

- bolting
- filter housing
- heat exchanger (shell)
- piping
- pump casing
- rupture disk
- strainer housing
- trap
- tubing
- valve body

The AE system component intended function within the scope of license renewal is to provide pressure boundary.

2.3.3.13P.2 Staff Evaluation

The staff reviewed LRA Section 2.3.3.13.2 and UFSAR Section 11.4 using the Tier-1 evaluation methodology described in SER Section 2.3 and the guidance described in SRP-LR Section 2.3.

In conducting its review, the staff evaluated the system functions described in the LRA and UFSAR to verify that the applicant has not omitted from the scope of license renewal any components with intended functions as required by 10 CFR 54.4(a). The staff then reviewed those components that the applicant has identified as within the scope of license renewal to verify that the applicant has not omitted any passive and long-lived components subject to an AMR as required by 10 CFR 54.21(a)(1).

2.3.3.13P.3 Conclusion

The staff reviewed the LRA to determine whether the applicant failed to identify any SSCs within the scope of license renewal or subject to an AMR. The staff finds no such omissions. On the basis of its review, the staff concludes that there is reasonable assurance that the applicant has adequately identified the AE system components within the scope of license renewal, as required by 10 CFR 54.4(a), and those subject to an AMR, as required by 10 CFR 54.21(a)(1).

2.3.3.13Q Building (Drainage System Components)

2.3.3.13Q.1 Summary of Technical Information in the Application

The building (BLD) system removes operational waste fluids from their points of origin in a controlled manner and delivers them to a suitable disposal system. The BLD system includes floor drains and the site sewers. This system classification also includes buildings and structures which are evaluated in LRA Section 2.4.

The failure of nonsafety-related SSCs in the BLD system potentially could prevent the satisfactory accomplishment of a safety-related function.

LRA Table 2.3.3-13-46 identifies the following BLD system component types within the scope of license renewal and subject to an AMR:

- bolting
- piping

The BLD component intended function within the scope of license renewal is to provide pressure boundary.

2.3.3.13Q.2 Staff Evaluation

The staff reviewed LRA Section 2.3.3.13.2 and UFSAR Section 10.16 using the Tier-1 evaluation methodology described in SER Section 2.3 and the guidance described in SRP-LR Section 2.3.

In conducting its review, the staff evaluated the system functions described in the LRA and UFSAR to verify that the applicant has not omitted from the scope of license renewal any components with intended functions as required by 10 CFR 54.4(a). The staff then reviewed those components that the applicant has identified as within the scope of license renewal to verify that the applicant has not omitted any passive and long-lived components subject to an AMR as required by 10 CFR 54.21(a)(1).

2.3.3.13Q.3 Conclusion

The staff reviewed the LRA to determine whether the applicant failed to identify any SSCs within the scope of license renewal or subject to an AMR. The staff finds no such omissions. On the basis of its review, the staff concludes that there is reasonable assurance that the applicant has adequately identified the BLD system components within the scope of license renewal, as required by 10 CFR 54.4(a), and those subject to an AMR, as required by 10 CFR 54.21(a)(1).

2.3.3.13R Circulating Water Priming

2.3.3.13R.1 Summary of Technical Information in the Application

The circulating water priming (CWP) system provides air evacuation from the discharge side of the main condenser. The system ensures that air will not hinder circulating water flow by collecting in the upper portions of the condenser water boxes or in the upper portion of the circulating water discharge piping.

The failure of nonsafety-related SSCs in the CWP system potentially could prevent the satisfactory accomplishment of a safety-related function.

LRA Table 2.3.3-13-47 identifies the following CWP system component types within the scope of license renewal and subject to an AMR:

- bolting
- piping
- pump casing
- tank
- trap
- tubing
- valve body

The CWP system component intended function within the scope of license renewal is to provide pressure boundary.

2.3.3.13R.2 Staff Evaluation

The staff reviewed LRA Section 2.3.3.13.2 and UFSAR Section 11.63 using the Tier-1 evaluation methodology described in SER Section 2.3 and the guidance described in SRP-LR Section 2.3.

In conducting its review, the staff evaluated the system functions described in the LRA and UFSAR to verify that the applicant has not omitted from the scope of license renewal any components with intended functions as required by 10 CFR 54.4(a). The staff then reviewed those components that the applicant has identified as within the scope of license renewal to verify that the applicant has not omitted any passive and long-lived components subject to an AMR as required by 10 CFR 54.21(a)(1).

2.3.3.13R.3 Conclusion

The staff reviewed the LRA to determine whether the applicant failed to identify any SSCs within the scope of license renewal or subject to an AMR. The staff finds no such omissions. On the basis of its review, the staff concludes that there is reasonable assurance that the applicant has adequately identified the CWP system components within the scope of license renewal, as required by 10 CFR 54.4(a), and those subject to an AMR, as required by 10 CFR 54.21(a)(1).

2.3.3.13S Extraction Steam

2.3.3.13S.1 Summary of Technical Information in the Application

The extraction steam (ES) system supplies steam to the shell side of various feedwater heaters for condensate and feedwater heating. Extraction steam is piped from the main turbine casing and cross-around piping to the shells of two parallel strings of reactor feedwater heaters.

The failure of nonsafety-related SSCs in the ES system potentially could prevent the satisfactory accomplishment of a safety-related function.

LRA Table 2.3.3-13-48 identifies the following ES system component types within the scope of license renewal and subject to an AMR:

- bolting
- expansion joint
- orifice
- piping
- tubing
- valve body

The ES system component intended function within the scope of license renewal is to provide pressure boundary.

2.3.3.13S.2 Staff Evaluation

The staff reviewed LRA Section 2.3.3.13.2 and UFSAR Section 11.5.4.3 using the Tier-1 evaluation methodology described in SER Section 2.3 and the guidance described in SRP-LR Section 2.3.

In conducting its review, the staff evaluated the system functions described in the LRA and UFSAR to verify that the applicant has not omitted from the scope of license renewal any components with intended functions as required by 10 CFR 54.4(a). The staff then reviewed those components that the applicant has identified as within the scope of license renewal to verify that the applicant has not omitted any passive and long-lived components subject to an AMR as required by 10 CFR 54.21(a)(1).

2.3.3.13S.3 Conclusion

The staff reviewed the LRA to determine whether the applicant failed to identify any SSCs within the scope of license renewal or subject to an AMR. The staff finds no such omissions. On the basis of its review, the staff concludes that there is reasonable assurance that the applicant has adequately identified the ES system components within the scope of license renewal, as required by 10 CFR 54.4(a), and those subject to an AMR, as required by 10 CFR 54.21(a)(1).

2.3.3.13T Heater Drain

2.3.3.13T.1 Summary of Technical Information in the Application

The heater drain (HD) system provides proper level and control for the moisture separator and feedwater heaters by providing drain capability to the main condenser. Condensate drainage from the drain coolers of each feedwater heater flows to the next lower pressure heater by means of pressure differential between successive heaters. Condensate flow may be aided by a heater drain pump between the two lowest pressure heaters in each string.

The failure of nonsafety-related SSCs in the HD system potentially could prevent the satisfactory accomplishment of a safety-related function.

LRA Table 2.3.3-13-49 identifies the following HD system component types within the scope of license renewal and subject to an AMR:

- bolting
- piping
- pump casing
- tank
- tubing
- valve body

The HD system component intended function within the scope of license renewal is to provide pressure boundary.

2.3.3.13T.2 Staff Evaluation

The staff reviewed LRA Section 2.3.3.13.2 and UFSAR Section 11.8.3.2 using the Tier-1 evaluation methodology described in SER Section 2.3 and the guidance described in SRP-LR Section 2.3.

In conducting its review, the staff evaluated the system functions described in the LRA and UFSAR to verify that the applicant has not omitted from the scope of license renewal any components with intended functions as required by 10 CFR 54.4(a). The staff then reviewed those components that the applicant has identified as within the scope of license renewal to verify that the applicant has not omitted any passive and long-lived components subject to an AMR as required by 10 CFR 54.21(a)(1).

2.3.3.13T.3 Conclusion

The staff reviewed the LRA to determine whether the applicant failed to identify any SSCs within the scope of license renewal or subject to an AMR. The staff finds no such omissions. On the basis of its review, the staff concludes that there is reasonable assurance that the applicant has adequately identified the HD system components within the scope of license renewal, as required by 10 CFR 54.4(a), and those subject to an AMR, as required by 10 CFR 54.21(a)(1).

2.3.3.13U Heater Vent

2.3.3.13U.1 Summary of Technical Information in the Application

The heater vent (HV) system provides venting of non-condensable gases back to the main condenser.

The failure of nonsafety-related SSCs in the HV system potentially could prevent the satisfactory accomplishment of a safety-related function.

LRA Table 2.3.3-13-50 identifies the following HV system component types within the scope of license renewal and subject to an AMR:

- bolting
- orifice
- piping
- tank
- tubing
- valve body

The HV system component intended function within the scope of license renewal is to provide pressure boundary.

2.3.3.13U.2 Staff Evaluation

The staff reviewed LRA Section 2.3.3.13.2, the Tier-1 evaluation methodology described in SER Section 2.3, and the guidance described in SRP-LR Section 2.3.

In conducting its review, the staff evaluated the system functions described in the LRA to verify that the applicant has not omitted from the scope of license renewal any components with intended functions as required by 10 CFR 54.4(a). The staff then reviewed those components that the applicant has identified as within the scope of license renewal to verify that the applicant has not omitted any passive and long-lived components subject to an AMR as required by 10 CFR 54.21(a)(1).

2.3.3.13U.3 Conclusion

The staff reviewed the LRA to determine whether the applicant failed to identify any SSCs within the scope of license renewal or subject to an AMR. The staff finds no such omissions. On the basis of its review, the staff concludes that there is reasonable assurance that the applicant has adequately identified the HV system components within the scope of license renewal, as required by 10 CFR 54.4(a), and those subject to an AMR, as required by 10 CFR 54.21(a)(1).

2.3.3.13V Make-up Demineralizer

2.3.3.13V.1 Summary of Technical Information in the Application

The make-up demineralizer (MUD) system provides a supply of treated water that may be used as make-up for the station and reactor cycles. The MUD system consists of one train that consists of a cation, anion, and a mixed bed ion exchanger.

The failure of nonsafety-related SSCs in the MUD system potentially could prevent the satisfactory accomplishment of a safety-related function.

LRA Table 2.3.3-13-53 identifies the following MUD system component types within the scope of license renewal and subject to an AMR:

- bolting
- filter housing
- piping
- pump casing
- tank
- tubing
- valve body

The MUD system component intended function within the scope of license renewal is to provide pressure boundary.

2.3.3.13V.2 Staff Evaluation

The staff reviewed LRA Section 2.3.3.13.2 and UFSAR Section 10.13 using the Tier-1 evaluation methodology described in SER Section 2.3 and the guidance described in SRP-LR Section 2.3.

In conducting its review, the staff evaluated the system functions described in the LRA and UFSAR to verify that the applicant has not omitted from the scope of license renewal any components with intended functions as required by 10 CFR 54.4(a). The staff then reviewed those components that the applicant has identified as within the scope of license renewal to verify that the applicant has not omitted any passive and long-lived components subject to an AMR as required by 10 CFR 54.21(a)(1).

2.3.3.13V.3 Conclusion

The staff reviewed the LRA to determine whether the applicant failed to identify any SSCs within the scope of license renewal or subject to an AMR. The staff finds no such omissions. On the basis of its review, the staff concludes that there is reasonable assurance that the applicant has adequately identified the MUD system components within the scope of license renewal, as required by 10 CFR 54.4(a), and those subject to an AMR, as required by 10 CFR 54.21(a)(1).

2.3.3.13W Seal Oil

2.3.3.13W.1 Summary of Technical Information in the Application

The seal oil (SO) system provides shaft sealing for the main generator.

The failure of nonsafety-related SSCs in the SO system potentially could prevent the satisfactory accomplishment of a safety-related function.

LRA Table 2.3.3-13-55 identifies the following SO system component types within the scope of license renewal and subject to an AMR:

- bolting
- filter housing
- piping
- pump casing
- sight glass
- tank
- tubing
- valve body

The SO system component intended function within the scope of license renewal is to provide pressure boundary.

2.3.3.13W.2 Staff Evaluation

The staff reviewed LRA Section 2.3.3.13.2 and UFSAR Section 11.23 using the Tier-1 evaluation methodology described in SER Section 2.3 and the guidance described in SRP-LR Section 2.3.

In conducting its review, the staff evaluated the system functions described in the LRA and UFSAR to verify that the applicant has not omitted from the scope of license renewal any components with intended functions as required by 10 CFR 54.4(a). The staff then reviewed those components that the applicant has identified as within the scope of license renewal to verify that the applicant has not omitted any passive and long-lived components subject to an AMR as required by 10 CFR 54.21(a)(1).

2.3.3.13W.3 Conclusion

The staff reviewed the LRA to determine whether the applicant failed to identify any SSCs within the scope of license renewal or subject to an AMR. The staff finds no such omissions. On the basis of its review, the staff concludes that there is reasonable assurance that the applicant has adequately identified the SO system components within the scope of license renewal, as required by 10 CFR 54.4(a), and those subject to an AMR, as required by 10 CFR 54.21(a)(1).

2.3.3.13X Turbine Building Closed Cooling Water

2.3.3.13X.1 Summary of Technical Information in the Application

The turbine building closed cooling water (TBCCW) system supplies demineralized water to cool various nonsafety-related auxiliary equipment located in the turbine building in support of power generation. The system consists of two pumps, two 100 percent capacity heat exchangers and the necessary controls, piping, and instrumentation. Station service water provides the cooling medium for the TBCCW heat exchangers, however, it is automatically isolated if service water pressure drops to a preset value which could occur under a condition of concurrent loss-of-coolant accident and loss of offsite power. No essential equipment is cooled by the TBCCW system.

The failure of nonsafety-related SSCs in the TBCCW system potentially could prevent the satisfactory accomplishment of a safety-related function.

LRA Table 2.3.3-13-56 identifies the following TBCCW system component types within the scope of license renewal and subject to an AMR:

- bolting
- heat exchanger (shell)
- piping
- pump casing
- tank
- tubing
- valve body

The TBCCW system component intended function within the scope of license renewal is to provide pressure boundary.

2.3.3.13X.2 Staff Evaluation

The staff reviewed LRA Section 2.3.3.13.2 and UFSAR Section 10.10 using the Tier-1 evaluation methodology described in SER Section 2.3 and the guidance described in SRP-LR Section 2.3.

In conducting its review, the staff evaluated the system functions described in the LRA and UFSAR to verify that the applicant has not omitted from the scope of license renewal any components with intended functions as required by 10 CFR 54.4(a). The staff then reviewed those components that the applicant has identified as within the scope of license renewal to verify that the applicant has not omitted any passive and long-lived components subject to an AMR as required by 10 CFR 54.21(a)(1).

2.3.3.13X.3 Conclusion

The staff reviewed the LRA to determine whether the applicant failed to identify any SSCs within the scope of license renewal or subject to an AMR. The staff finds no such omissions. On the basis of its review, the staff concludes that there is reasonable assurance that the applicant has adequately identified the TBCCW system components within the scope of license renewal, as required by 10 CFR 54.4(a), and those subject to an AMR, as required by 10 CFR 54.21(a)(1).

2.3.3.13Y Main Turbine Generator

2.3.3.13Y.1 Summary of Technical Information in the Application

The main turbine generator (TG) system converts the thermodynamic energy of steam into electrical energy for use on the transmission network and the station auxiliary busses.

The failure of nonsafety-related SSCs in the TG system potentially could prevent the satisfactory accomplishment of a safety-related function.

LRA Table 2.3.3-13-57 identifies the following TG system component types within the scope of license renewal and subject to an AMR:

- bolting
- filter housing
- piping
- pump casing
- turbine casing
- tubing
- valve body

The TG system component intended function within the scope of license renewal is to provide pressure boundary.

2.3.3.13Y.2 Staff Evaluation

The staff reviewed LRA Section 2.3.3.13.2 and UFSAR Section 11.2 using the Tier-1 evaluation methodology described in SER Section 2.3 and the guidance described in SRP-LR Section 2.3.

In conducting its review, the staff evaluated the system functions described in the LRA and UFSAR to verify that the applicant has not omitted from the scope of license renewal any components with intended functions as required by 10 CFR 54.4(a). The staff then reviewed those components that the applicant has identified as within the scope of license renewal to verify that the applicant has not omitted any passive and long-lived components subject to an AMR as required by 10 CFR 54.21(a)(1).

2.3.3.13Y.3 Conclusion

The staff reviewed the LRA to determine whether the applicant failed to identify any SSCs within the scope of license renewal or subject to an AMR. The staff finds no such omissions. On the basis of its review, the staff concludes that there is reasonable assurance that the applicant has adequately identified the TG system components within the scope of license renewal, as required by 10 CFR 54.4(a), and those subject to an AMR, as required by 10 CFR 54.21(a)(1).

2.3.3.13Z Turbine Lube Oil

2.3.3.13Z.1 Summary of Technical Information in the Application

The turbine lube oil (TLO) system provides lube oil for lubrication of the main turbine.

The failure of nonsafety-related SSCs in the TLO system potentially could prevent the satisfactory accomplishment of a safety-related function.

LRA Table 2.3.3-13-58 identifies the following TLO system component types within the scope of license renewal and subject to an AMR:

- bolting
- filter housing
- heat exchanger (shell)
- piping
- pump casing
- strainer housing
- tank
- tubing
- valve body

The TLO system component intended function within the scope of license renewal is to provide pressure boundary.

2.3.3.13Z.2 Staff Evaluation

The staff reviewed LRA Section 2.3.3.13.2 and UFSAR Section 11.2.3 using the Tier-1 evaluation methodology described in SER Section 2.3 and the guidance described in SRP-LR Section 2.3.

In conducting its review, the staff evaluated the system functions described in the LRA and UFSAR to verify that the applicant has not omitted from the scope of license renewal any components with intended functions as required by 10 CFR 54.4(a). The staff then reviewed those components that the applicant has identified as within the scope of license renewal to verify that the applicant has not omitted any passive and long-lived components subject to an AMR as required by 10 CFR 54.21(a)(1).

2.3.3.13Z.3 Conclusion

The staff reviewed the LRA to determine whether the applicant failed to identify any SSCs within the scope of license renewal or subject to an AMR. The staff finds no such omissions. On the basis of its review, the staff concludes that there is reasonable assurance that the applicant has adequately identified the TLO system components within the scope of license renewal, as required by 10 CFR 54.4(a), and those subject to an AMR, as required by 10 CFR 54.21(a)(1).

2.3.3.13AA Hydrogen Water Chemistry

2.3.3.13AA.1 Summary of Technical Information in the Application

The hydrogen water chemistry (HWC) system mitigates the chemical conditions that allow IGSCC in the recirculation piping and reactor vessels internals. The HWC system injects hydrogen into the reactor feedwater at the suction of the feedwater pumps.

The failure of nonsafety-related SSCs in the HWC system potentially could prevent the satisfactory accomplishment of a safety-related function.

LRA Table 2.3.3-13-51 identifies the following HWC system component types within the scope of license renewal and subject to an AMR:

- bolting
- piping
- tubing
- valve body

The HWC system component intended function within the scope of license renewal is to provide pressure boundary.

2.3.3.13AA.2 Staff Evaluation

The staff reviewed LRA Section 2.3.3.13.2 and UFSAR Sections 4.2.5, 11.8.3.1 and K.4.7. using the evaluation methodology described in SER Section 2.3 and the guidance described in SRP-LR Section 2.3.

In conducting its review, the staff evaluated the system functions described in the LRA and UFSAR to verify that the applicant has not omitted from the scope of license renewal any components with intended functions as required by 10 CFR 54.4(a). The staff then reviewed those components that the applicant has identified as within the scope of license renewal to verify that the applicant has not omitted any passive and long-lived components subject to an AMR as required by 10 CFR 54.21(a)(1).

2.3.3.13AA.3 Conclusion

The staff reviewed the LRA to determine whether the applicant failed to identify any SSCs within the scope of license renewal or subject to an AMR. The staff finds no such omissions. On the basis of its review, the staff concludes that there is reasonable assurance that the applicant has adequately identified the HWC system components within the scope of license renewal, as required by 10 CFR 54.4(a), and those subject to an AMR, as required by 10 CFR 54.21(a)(1).

The remaining systems shown in LRA Table 2.3.3.13-A as within the scope of license renewal with potential for physical interaction with safety-related components are addressed elsewhere in other LRA sections listed here:

- 2.3.1   CRD
- 2.3.1   HCUs
- 2.3.1   NB
- 2.3.2.1   RHR
- 2.3.2.2   CS
- 2.3.2.4   HPCI
- 2.3.2.5   CST
- 2.3.2.5   RCIC
- 2.3.2.6   SBGT
- 2.3.3.1   SLC
- 2.3.3.2   SW
- 2.3.3.2   RHRSW
- 2.3.3.3   RBCCW
- 2.3.3.4   DG and auxiliaries
- 2.3.3.4   DLO
- 2.3.3.5   FPC
- 2.3.3.5   FPC filter demineralizer
- 2.3.3.5   SBFPC
- 2.3.3.6   FO
- 2.3.3.7   IA
- 2.3.3.7 N_2
- 2.3.3.8   fire protection
- 2.3.3.10   HB
- 2.3.3.10   HVAC
- 2.3.3.11   containment air dilution
- 2.3.3.11   PASS
- 2.3.3.11   PCAC
- 2.3.4.2   condensate

2.3.4 Steam and Power Conversion Systems

In LRA Section 2.3.4, the applicant identified the SCs of the steam and power conversion systems that are subject to an AMR for license renewal.

The applicant described the supporting SCs of the steam and power conversion systems in the following LRA Sections:

- 2.3.4.1   auxiliary steam
- 2.3.4.2   condensate
- 2.3.4.3   main steam
- 2.3.4.4   101 (main steam, extraction steam, and auxiliary steam instruments)

The staff's review findings regarding LRA Sections 2.3.4.1 – 2.3.4.4 are presented in SER Sections 2.3.4.1 – 2.3.4.4, respectively.

2.3.4.1 Auxiliary Steam

2.3.4.1.1 Summary of Technical Information in the Application

LRA Section 2.3.4.1 describes the auxiliary steam (AS) system, which provides steam from MS piping to the steam jet air ejector to maintain main condenser vacuum. The AS system consists of the steam jet air ejector and associated equipment.

The failure of nonsafety-related SSCs in the AS system potentially could prevent the satisfactory accomplishment of a safety-related function.

LRA Table 2.3.4-1 identifies the component types in the main condenser and MSIV leakage pathway that supports the 10 CFR 54.4(a)(2) intended function of the AS system and LRA Table 2.3.3-13-45 identifies the AS system component types within the scope of license renewal and subject to an AMR:

- bolting
- condenser
- expansion joint
- heat exchanger (shell)
- heat exchanger (tubes)
- piping
- orifice
- strainer housing
- steam trap
- thermowell
- tubing
- valve body

The AS system component intended functions within the scope of license renewal include the following:

- pressure boundary
- holdup and plateout of fission products

2.3.4.1.2 Staff Evaluation

The staff reviewed LRA Section 2.3.4.1 and UFSAR Section 11.4 using the Tier-2 evaluation methodology described in SER Section 2.3 and the guidance in SRP-LR Section 2.3.

In conducting its review, the staff evaluated the system functions described in the LRA and UFSAR to verify that the applicant has not omitted from the scope of license renewal any components with intended functions as required by 10 CFR 54.4(a). The staff then reviewed those components that the applicant has identified as within the scope of license renewal to verify that the applicant has not omitted any passive and long-lived components subject to an AMR as required by 10 CFR 54.21(a)(1).

2.3.4.1.3 Conclusion

The staff reviewed the LRA and accompanying license renewal drawings to determine whether the applicant failed to identify any SSCs within the scope of license renewal or subject to an AMR. The staff finds no such omissions. On the basis of its review, the staff concludes that there is reasonable assurance that the applicant has adequately identified the AS system components within the scope of license renewal, as required by 10 CFR 54.4(a), and those subject to an AMR, as required by 10 CFR 54.21(a)(1).

2.3.4.2 Condensate

2.3.4.2.1 Summary of Technical Information in the Application

LRA Section 2.3.4.2 describes the condensate system, which receives condensed steam from the condenser and supplies it to the reactor feedwater system as well as such other components and systems as the air ejector condensers, steam packing exhausters, and CRD pumps. The condensate system consists of a single train with three parallel pumps drawing condensate from the two main condenser hotwells and includes the main condenser. During normal operation, all three pumps provide sufficient condensate flow capacity and net positive suction head to the reactor feedwater pumps during full power operation. Condensate flow to the reactor feed pumps passes through two parallel low-pressure feedwater heater strings, each with three heaters. Condensate flow exiting the low-pressure heaters is provided to a common reactor feed pump suction header.

The failure of nonsafety-related SSCs in the condensate system potentially could prevent the satisfactory accomplishment of a safety-related function.

LRA Tables 2.3.4-1 and 2.3.3-13-2 identify the following condensate component types within the scope of license renewal and subject to an AMR:

- bolting
- condenser
- expansion joint
- heat exchanger (shell)
- heat exchanger (tubes)
- orifice
- piping
- pump casing
- steam trap
- strainer housing
- tank
- thermowell
- tubing
- valve body

The condensate system component intended functions within the scope of license renewal include the following:

- pressure boundary
- holdup and plateout of fission products

2.3.4.2.2 Staff Evaluation

The staff reviewed LRA Sections 2.3.4.2 and 2.3.3.13, and UFSAR Section 11.8 using the Tier-2 evaluation methodology described in SER Section 2.3 and the guidance in SRP-LR Section 2.3.

In conducting its review, the staff evaluated the system functions described in the LRA and UFSAR to verify that the applicant has not omitted from the scope of license renewal any components with intended functions as required by 10 CFR 54.4(a). The staff then reviewed those components that the applicant has identified as within the scope of license renewal to verify that the applicant has not omitted any passive and long-lived components subject to an AMR as required by 10 CFR 54.21(a)(1).

2.3.4.2.3 Conclusion

The staff reviewed the LRA to determine whether the applicant failed to identify any SSCs within the scope of license renewal or subject to an AMR. The staff finds no such omissions. On the basis of its review, the staff concludes that there is reasonable assurance that the applicant has adequately identified the condensate system components within the scope of license renewal, as required by 10 CFR 54.4(a), and those subject to an AMR, as required by 10 CFR 54.21(a)(1).

2.3.4.3 Main Steam

2.3.4.3.1 Summary of Technical Information in the Application

LRA Section 2.3.4.3 describes the MS system, which completes the transmission of steam from the seismic Class I steam piping to the main turbine at a controlled pressure during normal operation. The MS system consists of nonsafety-related components. (The nuclear boiler system contains the seismic Class I portion of the MS system which extends from the reactor vessel to the restraint at the second MS isolation valve. The system consists of the non-seismic Class I components beyond this point.) The MS system includes the turbine stop and control valves. A low-point drain line is downstream of each turbine control valve continuously draining the steam line low points through an orificed header to the condenser hotwell. The MS system has the ability to bypass the turbine when necessary. The main turbine bypass system has two valve chests, each with five automatically operated regulating bypass valves proportionally controlled by the turbine pressure regulator and control system. The bypass system opens whenever the amount of steam admitted into the turbine is less than that generated by the reactor. The MS system provides main turbine sealing steam.

The failure of nonsafety-related SSCs in the MS system potentially could prevent the satisfactory accomplishment of a safety-related function.

LRA Tables 2.3.4-1and 2.3.3-13-52 identify the following MS system component types within the scope of license renewal and subject to an AMR:

- bolting
- condenser
- expansion joint
- heat exchanger (shell)
- heat exchanger (tubes)
- orifice
- piping
- steam trap
- strainer housing
- thermowell
- tubing
- valve body

The MS system component intended functions within the scope of license renewal include the following:

- pressure boundary
- holdup and plateout of fission products

2.3.4.3.2 Staff Evaluation

The staff reviewed LRA Section 2.3.4.3 and UFSAR Sections 11.4 and 11.5 using the Tier-2 evaluation methodology described in SER Section 2.3 and the guidance in SRP-LR Section 2.3.

In conducting its review, staff evaluated the system functions described in the LRA and UFSAR to verify that the applicant has not omitted from the scope of license renewal any components with intended functions as required by 10 CFR 54.4(a). The staff then reviewed those components that the applicant has identified as within the scope of license renewal to verify that the applicant has not omitted any passive and long-lived components subject to an AMR as required by 10 CFR 54.21(a)(1).

2.3.4.3.3 Conclusion

The staff reviewed the LRA and accompanying license renewal drawings to determine whether the applicant failed to identify any SSCs within the scope of license renewal or subject to an AMR. The staff finds no such omissions. On the basis of its review, the staff concludes that there is reasonable assurance that the applicant has adequately identified the MS system components within the scope of license renewal, as required by 10 CFR 54.4(a), and those subject to an AMR, as required by 10 CFR 54.21(a)(1).

2.3.4.4 101 (Main Steam, Extraction Steam, and Auxiliary Steam Instruments)

2.3.4.4.1 Summary of Technical Information in the Application

LRA Section 2.3.4.4 describes the 101 system (main steam, extraction steam, and auxiliary steam instruments), which provides indication, alarm, and control functions for its associated systems. This system code includes various instrumentation components for main steam, extraction steam, and auxiliary steam. Although the 101 system consists mainly of EIC components, certain mechanical instrumentation components are included as well.

The failure of nonsafety-related SSCs in the 101 system (main steam, extraction steam, and auxiliary steam instruments) potentially could prevent the satisfactory accomplishment of a safety-related function.

LRA Table 2.3.4-1 identifies the following 101 system (main steam, extraction steam, and auxiliary steam instruments) component types within the scope of license renewal and subject to an AMR:

- bolting
- condenser
- orifice
- expansion joint
- heat exchanger (tubes)
- piping
- strainer housing
- thermowell
- steam trap
- tubing
- valve body

The 101 (main steam, extraction steam, and auxiliary steam instruments) component intended functions within the scope of license renewal include the following:

- pressure boundary
- holdup and plateout of fission products

2.3.4.4.2 Staff Evaluation

The staff reviewed LRA Section 2.3.4.4 using the Tier-1 evaluation methodology described in SER Section 2.3 and the guidance in SRP-LR Section 2.3.

In conducting its review, the staff evaluated the system functions described in the LRA and UFSAR to verify that the applicant has not omitted from the scope of license renewal any components with intended functions as required by 10 CFR 54.4(a). The staff then reviewed those components that the applicant has identified as within the scope of license renewal to verify that the applicant has not omitted any passive and long-lived components subject to an AMR as required by 10 CFR 54.21(a)(1).

2.3.4.4.3 Conclusion

The staff reviewed the LRA and accompanying license renewal drawings to determine whether the applicant failed to identify any SSCs within the scope of license renewal or subject to an AMR. The staff finds no such omissions. On the basis of its review, the staff concludes that there is reasonable assurance that the applicant has adequately identified the 101 (main steam, extraction steam, and auxiliary steam instruments) components within the scope of license renewal, as required by 10 CFR 54.4(a), and those subject to an AMR, as required by 10 CFR 54.21(a)(1).

2.4 Scoping and Screening Results: Structures

This section documents the staff's review of the applicant's scoping and screening results for structures. Specifically, this section discusses:

- primary containment
- reactor building
- intake structure
- process facilities
- yard structures
- bulk commodities

In accordance with the requirements of 10 CFR 54.21(a)(1), the applicant must list passive, long-lived SCs within the scope of license renewal and subject to an AMR. To verify that the applicant properly implemented its methodology, the staff's review focused on the implementation results. This focus allowed the staff to confirm that there were no omissions of SCs that meet the scoping criteria and are subject to an AMR.

The staff's evaluation of the information in the LRA was the same for all structures. The objective was to determine whether the applicant has identified, in accordance with 10 CFR 54.4, components and supporting structures for structures that appear to meet the license renewal scoping criteria. Similarly, the staff evaluated the applicant's screening results to verify that all passive, long-lived components were subject to an AMR as required by 10 CFR 54.21(a)(1).

In its scoping evaluation, the staff reviewed the applicable LRA sections and component drawings, focusing on components that have not been identified as within the scope of license renewal. The staff reviewed relevant licensing basis documents, including the UFSAR, for each structure to determine whether the applicant has omitted from the scope of license renewal components with intended functions as required by 10 CFR 54.4(a). The staff also reviewed the licensing basis documents to determine whether the LRA specified all intended functions as required by 10 CFR 54.4(a). The staff requested additional information to resolve any omissions or discrepancies identified.

After its review of the scoping results, the staff evaluated the applicant's screening results. For those SCs with intended functions, the staff sought to determine whether: (1) the functions are performed with moving parts or a change in configuration or properties or (2) the SCs are subject to replacement after a qualified life or specified time period, as required by

10 CFR 54.21(a)(1). For those meeting neither of these criteria, the staff sought to confirm that these SCs were subject to an AMR, as required by 10 CFR 54.21(a)(1). The staff requested additional information to resolve any omissions or discrepancies identified.

2.4.1 Primary Containment

2.4.1.1 Summary of Technical Information in the Application

LRA Section 2.4.1 describes the primary containment, which limits the release of fission products in postulated design basis accidents so offsite doses do not exceed the values specified in 10 CFR 50.67. Located inside the reactor building, the primary containment is a General Electric Mark I containment with a drywell (which encloses the reactor vessel and recirculation system), a pressure suppression chamber (commonly known as the torus), and a connecting vent system. When operating at power, the containment is flooded with N_2 to preclude the availability of oxygen. The drywell surrounds the reactor vessel and primary systems. The torus, containing water, is below the drywell and the vent system connecting it to the drywell terminates below the water surface. Access to the drywell is by its steel drywell head and personnel hatch as well as a double door air lock, equipment hatch, and one CRD access hatch. Access to the torus is by two personnel hatches. The primary containment components include the drywell, the torus, the reactor vessel and drywell bellows, and the shield wall. The drywell is a carbon steel structure that houses the reactor pressure vessel and its components. A reinforced concrete support structure, founded on bedrock, is part of the drywell support system. The torus is a toroid-shaped carbon steel pressure vessel below and encircling the drywell. The reactor vessel refueling bulkhead has two stainless steel bellows with backing plates, spring seals, and removable guard rings. The drywell to reactor building bellows assembly is similar to that of the reactor vessel refueling bulkhead. The shield wall (also known as the sacrificial shield wall) is a high-density, steel-reinforced, concrete cylindrical structure surrounding the vessel. The concrete is contained by inner and outer steel liner plates that also attach various system supports. The sacrificial shield wall provides lateral support for the reactor vessel to accommodate both seismic forces and jet forces from the breakage of any pipe attached to the vessel.

The primary containment has safety-related components relied upon to remain functional during and following DBEs. The failure of nonsafety-related primary containment SSCs potentially could prevent the satisfactory accomplishment of a safety-related function. In addition, the primary containment performs functions that support fire protection.

LRA Table 2.4-1 identifies the following primary containment component types within the scope of license renewal and subject to an AMR:

- steel and other metals
- concrete
- elastomers and other materials
- fluoropolymers and lubrite sliding surfaces

The primary containment component intended functions within the scope of license renewal include the following:

- shelter or protection to safety-related equipment, including radiation shielding and pipe whip restraint
- protective barrier for flood events
- heat sink during SBO or DBAs
- missile barrier
- pressure boundary
- structural or functional support for safety-related equipment

2.4.1.2 Staff Evaluation

The staff reviewed LRA Section 2.4.1 and UFSAR Sections 5.1.2 and 5.2 using the evaluation methodology described in SER Section 2.4 and the guidance in SRP-LR Section 2.4, "Scoping and Screening Results: Structures."

The staff evaluated the structural component functions described in the LRA and UFSAR to verify that the applicant has not omitted from the scope of license renewal any components with intended functions as required by 10 CFR 54.4(a). The staff then reviewed those components that the applicant has identified as within the scope of license renewal to verify that the applicant has not omitted any passive and long-lived components subject to an AMR as required by 10 CFR 54.21(a)(1).

2.4.1.3 Conclusion

The staff reviewed the LRA and related structural components to determine whether the applicant failed to identify any SSCs within the scope of license renewal or subject to an AMR. The staff finds no such omissions. On the basis of its review, the staff concludes that there is reasonable assurance that the applicant has adequately identified the primary containment components within the scope of license renewal, as required by 10 CFR 54.4(a), and those subject to an AMR, as required by 10 CFR 54.21(a)(1).

2.4.2 Reactor Building

2.4.2.1 Summary of Technical Information in the Application

LRA Section 2.4.2 describes the reactor building, which in design basis accidents contains leakage of airborne fission products to the environment within the dose limits specified in 10 CFR 50.67 and supports and protects the reactor and its systems. The reactor building completely encloses the primary containment and houses the refueling and reactor servicing equipment (platforms and cranes), new and spent fuel storage facilities, reactor core isolation cooling system, SBGT system, reactor cleanup demineralizer system, SLC system, CRD system equipment, reactor core and containment cooling systems, and electrical equipment components. The seismic Class I reactor building is constructed of monolithic reinforced concrete floors and walls up to the refueling level and of steel framing covered by insulated sealed siding and roof decking above. The siding and roofing can withstand limited internal overpressure before it is relieved by venting through blowout panels. A biological shield wall, part of the reactor building, encircles the primary containment, protects the containment vessel and the reactor system against potential external missiles, and shields personnel to reduce dose.

The reactor building bridge crane, which services the reactor and the refueling area, is designed seismic Class II with supports designed seismic Class I. The crane bridge and trolley wheels have seismic holddown lugs for crane stability in a hypothetical maximum earthquake. The new fuel storage vault, part of the seismic Class I reactor building, houses new fuel storage racks, each designed as seismic Class I while loaded with fuel. The spent fuel storage pool in the reactor building is lined with stainless steel. The pool liner is seam-welded ASTM-A240 Type 304 stainless steel with pipe sleeves welded to both sides of the liner plate. The spent fuel storage racks are assemblies of individual storage cells consisting of Type 304L stainless steel boxes welded together. The seismic Class I refueling platform, the principal means of transporting fuel assemblies back and forth, travels on tracks extending along each side between the reactor well and the storage pool.

The reactor building has safety-related components relied upon to remain functional during and following DBEs. The failure of nonsafety-related reactor building SSCs potentially could prevent the satisfactory accomplishment of a safety-related function. In addition, the reactor building performs functions that support fire protection, ATWS, and SBO.

LRA Table 2.4-2 identifies the following reactor building component types within the scope of license renewal and subject to an AMR:

- steel and other metals
- concrete

The reactor building component intended functions within the scope of license renewal include the following:

- shelter or protection to safety-related equipment, including radiation shielding and pipe whip restraint

- rated fire barrier to confine or retard a fire from spreading

- protective barrier for flood events

- missile barrier

- pressure boundary

- structural or functional support to nonsafety-related equipment the failure of which could impact safety-related equipment

- structural or functional support for safety-related equipment

2.4.2.2 Staff Evaluation

The staff reviewed LRA Section 2.4.2 and UFSAR Sections 5.3, 10.4, and 12.2.2 using the evaluation methodology described in SER Section 2.4 and the guidance in SRP-LR Section 2.4.

The staff evaluated the structural component functions described in the LRA and UFSAR to verify that the applicant has not omitted from the scope of license renewal any components with intended functions as required by 10 CFR 54.4(a). The staff then reviewed those components that the applicant has identified as within the scope of license renewal to verify that the applicant has not omitted any passive and long-lived components subject to an AMR as required by 10 CFR 54.21(a)(1).

2.4.2.3 Conclusion

The staff reviewed the LRA and related structural components to determine whether the applicant failed to identify any SSCs within the scope of license renewal or subject to an AMR. The staff finds no such omissions. On the basis of its review, the staff concludes that there is reasonable assurance that the applicant has adequately identified the reactor building components within the scope of license renewal, as required by 10 CFR 54.4(a), and those subject to an AMR, as required by 10 CFR 54.21(a)(1).

2.4.3 Intake Structure

2.4.3.1 Summary of Technical Information in the Application

LRA Section 2.4.3 describes the intake structure, which supports and protects equipment that draws water from the intake canal, located east of the station on the riverbank and divided into two rooms: the SW pump room (which also contains the diesel and electric fire pumps) and the circulating water pump room. The room housing the SW pumps is seismic Class I; the other is seismic Class II. The reinforced concrete and steel intake structure is founded entirely on bedrock. It has three pump bays for the vertical circulating water pumps, two SW bays for four SW pumps and two fire water pumps, three roller gates, and one sluice gate. Recirculation of warm discharge water by a concrete pipe connecting the discharge structure to the intake structure keeps the intake bays and SW bays free of ice. All bays have trash racks and stop log guides, traveling screens, and fine screen guides. Interconnection of the three pump bays is by removal of stop logs in center walls.

The intake structure has safety-related components relied upon to remain functional during and following DBEs. The failure of nonsafety-related intake structure SSCs potentially could prevent the satisfactory accomplishment of a safety-related function. In addition, the intake structure performs functions that support fire protection.

LRA Table 2.4-3 identifies the following intake structure component types within the scope of license renewal and subject to an AMR:

- steel and other metals
- concrete

The intake structure component intended functions within the scope of license renewal include the following:

- shelter or protection to safety-related equipment, including radiation shielding and pipe whip restraint

- rated fire barrier to confine or retard a fire from spreading

- protective barrier for flood events

- missile barrier

- structural or functional support to nonsafety-related equipment the failure of which could impact safety-related equipment

- structural or functional support for equipment required to meet fire protection, environmental qualification, pressurized thermal shock (PTS), ATWS, or SBO regulations

- structural or functional support for safety-related equipment

2.4.3.2 Staff Evaluation

The staff reviewed LRA Section 2.4.3 and UFSAR Sections 10.6.5, 10.11.3, and 12.2.6 using the evaluation methodology described in SER Section 2.4 and the guidance in SRP-LR Section 2.4.

The staff evaluated the structural component functions described in the LRA and UFSAR to verify that the applicant has not omitted from the scope of license renewal any components with intended functions as required by 10 CFR 54.4(a). The staff then reviewed those components that the applicant has identified as within the scope of license renewal to verify that the applicant has not omitted any passive and long-lived components subject to an AMR as required by 10 CFR 54.21(a)(1).

In RAI 2.4.3-1 dated August 3, 2006, the staff stated that Table 2.4.3 does not include the sluice gate, roller gates, trash racks, stop log guides, traveling screens, and fine screen guides within the intake structure, and the concrete pipe that connects the intake structure to the discharge structure. The staff requested that the applicant provide justification for not including them within the scope of license renewal.

In its response dated September 5, 2006, the applicant provided the following response:

Sluice gates and roller gates

The roller gates isolate the circulating water bays from the river and have no license renewal intended function. The sluice gate is used for de-icing. De-icing supports normal plant operation and is not credited for emergency operation, since warm circulating water flow would not be available with a loss of offsite power. The gates have no license renewal intended function and are not included in LRA Table 2.4-3.

Trash racks and traveling screens

The trash racks and traveling screens remove debris from the circulating and SW system flow path to prevent plugging of the condenser water box inlets and loss of SW flow. The circulating water bays and the SW bays have separate flow paths sharing a common wall. The trash racks prevent the high circulating water velocity from drawing large debris into the circulating water bays during normal plant operation. However, during emergency operations, the circulating water pumps are unnecessary and, in fact, may be unavailable due to a loss of offsite power. For normal and emergency operations, the SW pumps draw a much lower volume of water through the SW bays. The lower flow rates of the SW system are insufficient to transport large debris that could prevent the traveling screens from passing adequate flow to the SW pumps to allow for safe shutdown. Therefore, trash racks do not provide a license renewal intended function as required by 10 CFR 54.4(a)(1), (a)(2) or (a)(3).

The structural supports for the traveling screens are part of the screen-house structure, which is within the scope of license renewal and subject to an AMR. The traveling screens themselves perform their function with moving parts and a change in configuration and are therefore, not subject to an AMR in accordance with 10 CFR 54.21 (a)(I)(i), and are not included in LRA Table 2.4-3.

Stop log guides and fine screen guides

The stop log guides and fine screen guides do not perform a license renewal intended function. The purpose of the stop log guides is to hold temporary stop logs in place to allow inspections or maintenance. The fine screen guides do not perform a license renewal intended function because a fine screen is not utilized at VYNPS. Therefore, the stop log and fine screen guides do not provide a license renewal intended function as required by 10 CFR 54.4(a)(1), (a)(2) or (a)(3).

Concrete pipe

The concrete pipe connecting the intake structure to the discharge structure provides recirculation of warm condenser circulating water to keep the circulating water intake bays and SW bays free of ice. De-icing supports normal plant operation and is not credited for emergency operation, since warm circulating water flow would not be available with a loss of offsite power. Therefore, the concrete pipe does not provide a license renewal intended function as required by 10 CFR 54.4(a)(1), (a)(2) or (a)(3).

Based on its review, the staff finds the applicant's response to RAI 2.4.3-1 acceptable because the applicant has provided sufficient explanations for the function of the sluice gate, roller gates, trash racks, stop log guides, traveling screens and fine screen guides within the intake structure, and the concrete pipe that connects the intake structure to the discharge structure, and the basis of their exclusion from the license renewal intended function requirements of 10 CFR 54.4(a)(1), (2) or (3). The staff's concern described in RAI 2.4.3-1 is resolved.

2.4.3.3 Conclusion

The staff reviewed the LRA and related structural components to determine whether the applicant failed to identify any SSCs within the scope of license renewal or subject to an AMR. The staff finds no such omissions. On the basis of its review, the staff concludes that there is reasonable assurance that the applicant has adequately identified the intake structure components within the scope of license renewal, as required by 10 CFR 54.4(a), and those subject to an AMR, as required by 10 CFR 54.21(a)(1).

2.4.4 Process Facilities

2.4.4.1 Summary of Technical Information in the Application

LRA Section 2.4.4 describes the process facilities, buildings or structures designated as either seismic Class I or II for power generation and supporting processes with concrete floor slabs, structural steel floors, and platforms as required supported by concrete or structural steel columns, base slabs, and walls. Process facilities include alternate cooling cells and the cooling tower No. 2 deep basin, the control building, the plant stack, and the turbine building. Alternate cooling cell No. 2-1 and the cooling tower No. 2 deep basin provide a heat sink to remove decay heat and sensible heat from the primary system so the reactor can be shut down safely when the SW pumps are not available. Alternate cooling cell No. 2-1, adjoining cooling cell 2-2, and the cooling tower No. 2 deep basin, support and protect structures necessary for the heat sink.

The control building houses instrumentation and switches required for station operation with major instrumentation in the main control room. The cable vault and east and west switchgear rooms occupy the lower levels of the building. The plant stack (or main stack) discharges gases to the atmosphere from portions of the turbine building, reactor building, RDW building, SBGT system, and advanced off-gas system. The height of the stack ensures an elevated release and an enclosure at its base contains monitoring equipment. The turbine building houses the TG and auxiliaries including the condensate, feedwater, DG, and water treatment systems. Portions of the turbine building support and protect the EDGs and FO day tank areas.

The process facilities have safety-related components relied upon to remain functional during and following DBEs. The failure of nonsafety-related process facility SSCs potentially could prevent the satisfactory accomplishment of a safety-related function. In addition, the process facilities perform functions that support fire protection.

LRA Table 2.4-4 identifies the following process facilities component types within the scope of license renewal and subject to an AMR:

- steel and other metals
- concrete
- elastomer and other materials

The process facilities component intended functions within the scope of license renewal include the following:

- shelter or protection to safety-related equipment, including radiation shielding and pipe whip restraint

- rated fire barrier to confine or retard a fire from spreading

- protective barrier for flood events

- heat sink during SBO or DBAs

- missile barrier

- pressure boundary

- structural or functional support to nonsafety-related equipment the failure of which could impact safety-related equipment

- structural or functional support for equipment required to meet fire protection, environmental qualification, PTS, ATWS, or SBO regulations

- structural or functional support for safety-related equipment

2.4.4.2 Staff Evaluation

The staff reviewed LRA Section 2.4.4 and UFSAR Sections 10.8, 11.9, 12.2.3, 12.2.4, 12.2.5, and 12.2.6.4 using the evaluation methodology described in SER Section 2.4 and the guidance in SRP-LR Section 2.4.

The staff evaluated the structural component functions described in the LRA and UFSAR to verify that the applicant has not omitted from the scope of license renewal any components with intended functions as required by 10 CFR 54.4(a). The staff then reviewed those components that the applicant has identified as within the scope of license renewal to verify that the applicant has not omitted any passive and long-lived components subject to an AMR as required by 10 CFR 54.21(a)(1).

In RAI 2.4.4-1 dated August 3, 2006, the staff stated that Table 2.4-4 lists "Structural steel" as a component, and "Structural steel: beams, columns, plates " as another component. The staff requested that the applicant provide clarification for the two different components.

In its response dated September 5, 2006, the applicant provided the following response:

Table 2.4.4 lists these two different components.

"Structural steel: beams, columns, plates" is defined as:

- substructure or superstructure steel that is part of the primary structural support function of a building or structure, such as structural columns, support girders, beams, plates, connections, roofing joists, purlins, and wind bracing.

"Structural steel" is defined as:

- steel which does not perform a primary structural integrity function for a building but does provide secondary structural support for equipment or components within the building, or it may provide protection around openings in floors or walls and metal decking on the bottom of reinforced concrete floor slabs. Structural steel includes items such as grating, grating supports, embedded channels, angles, frames, and embedded inserts such as Unistrut™.

Based on its review, the staff finds the applicant's response to RAI 2.4.4-1 acceptable because it distinguishes the primary structural support function from a secondary structural support function of steel members. The staff's concern described in RAI 2.4.4-1 is resolved.

In Table 2.4-4, cooling tower cell No. 2-1, cooling tower cell No. 2-2, and foundation (cooling tower No. 2 deep basin) are listed as subject to an aging management review. On August 21, 2007, a portion of cooling tower cell No. 2-4 collapsed. The staff required verification as to whether the affected cells should be in the scope of license renewal and whether scoping for license renewal has been appropriately conducted with respect to the cooling towers.

In RAI 2.4.4-2 dated August 29, 2007, the staff requested that the applicant provide the results of the review performed to determine the impact of the circulating water piping, pipe supports, and west cooling tower cell (2-4) failures on license renewal scoping, screening, and applicable aging management programs. The staff also requested the applicant to include the following information:

A. A conclusion and basis as to whether the scoping results documented in the LRA, which initially determined that 9 of the11 west cooling tower cells were not within the scope of license renewal, are still valid.

B. If found that the west cooling tower cells (2-3 through 2-11) are within the scope of license renewal, provide the following:

  I. The potential effect of a circulating water piping, pipe supports, or structural failure of the nonsafety-related west cooling tower cells (2-3 through 2-11), which were not included within the scope of license renewal, on safety-related systems, structures, and components (in accordance with 10 CFR 54.4(a)(2)). Include the potential effect of debris entering the deep basin beneath the cooling tower.

  II. The details of any age related degradation which caused the failure of the circulating water piping, pipe supports, and west cooling tower cell. Include the results of the piping and pipe supports inspection related to the current failure and any previously performed, and a description of the identified aging mechanism(s).

C. Any impact on the aging management programs for circulating water piping, pipe supports, or cooling tower cells.

In letters dated September 27 and October 18, 2007, the applicant provided the following response:

Cooling Tower Background Information

VYNPS utilizes once-through condenser cooling from the Connecticut River supplemented by two forced draft cooling towers. Each tower consists of eleven cells, each cell equipped with its own forced draft fan. One cell in the west cooling tower, CT 2-1, provides a safety-related function as the heat sink for the Residual Heat Removal Service Water system (RHRSW) in the Alternate Cooling System (ACS) mode and is constructed as a Seismic Class I structure. The adjacent cell, CT 2-2, is also designed and constructed as a Seismic Class I structure to prevent adversely impacting the structural integrity of CT 2-1 during a seismic event.

CT 2-1 and CT 2-2 structures have similar construction as the other cooling tower cells for dead weight loads, but a more robust bracing system to withstand wind and seismic loading. They are constructed from high quality timber and use stainless steel hardware for all bolted connections. The structural columns were refurbished during the 1980's, followed by end wall refurbishment between 2002 and 2007. As required for activities associated with any safety-related and Seismic Class I systems, structures, and components (SSCs), the inspections and repairs on cooling tower cells CT 2-1 and CT 2-2 receive additional oversight by the site Engineering, Maintenance, and Quality Assurance (QA) groups.

- Different design. Safety-related Cell CT 2-1 and Seismic Class I Cell CT 2-2 design includes additional 4"x4" cross-bracing to withstand wind and seismic loading. In CT 2-1, some of the additional bracing is heavier 4" x 6" material.

- Different material specifications. Hardware for CT 2-1 and CT 2-2 is stainless steel, while the other towers may use carbon or galvanized steel. The stainless steel hardware minimizes potential iron salt attack at the bolted structural connections.

- Different level of quality. CT 2-1 and CT 2-2 are subject to the higher levels of oversight afforded to safety-related and Seismic Class I structures. The higher level of quality results in application of the station corrective action program to evaluate deficiencies and effect appropriate corrective actions.

- Different maintenance history. Because of their safety significance and higher level of quality, CT 2-1 and CT 2-2 have had more refurbishment during the past ten years than the other tower cells. During this period, the end wall of CT 2-1 and the partition walls of CT 2-1 and CT 2-2 have been replaced, including the vertical columns and structural hardware. The original end walls and partition walls remain in many of the non-Seismic Class I cells.

Response to Part A:

Cooling tower cells CT 2-1 and CT 2-2 are the only cells in the scope of license renewal. Failures of the other cells will not prevent satisfactory accomplishment of a safety function identified in 10 CFR 54.4(a)(1). The scoping results documented in the LRA remain valid. See the response to part B for further discussion of potential failures.

Cooling tower cell CT 2-1, which is part of the circulating water system, has the 10 CFR 54.4(a)(1) and (a)(3) intended function to support operation of the alternate cooling system by providing an alternate means of heat removal in the unlikely event that the service water pumps become inoperable. Therefore, CT 2-1 is in the scope of license renewal and subject to aging management review. Cell CT 2-1 itself and associated components of the residual heat removal service water (RHRSW) system fulfill the intended function. The credited RHRSW system components in CT 2-1 are the 24" carbon steel suction piping located in the RHRSW suction pit and the 16" and 20" carbon steel distribution piping that discharges water into the cooling tower from the RHRSW pumps. Aging management review results for RHRSW system components at CT 2-1 are provided in LRA Table 3.3.2-2. Circulating water piping is not relied on to perform the license renewal intended function of supporting alternate cooling system operation. The circulating water system piping has no other system intended functions in scope for 54.4(a)(1) or (a)(3). It does have a 54.4(a)(2) intended function to maintain integrity of nonsafety-related components such that no physical interaction with safety-related components could prevent satisfactory accomplishment of a safety function.

Response to Part B

Subpart I:

As indicated in the LRA and in response to Part A, west cooling tower cells CT 2-1 and CT 2-2 are within the scope of license renewal. The failure of cooling tower cell CT 2-4 or any other of the cooling tower cells, along with the associated circulating water piping and pipe supports, has no impact on the ability of the in-scope cooling tower cells and the Cooling Tower No. 2 (west cooling tower) deep basin to accomplish safety functions under design basis conditions. Cooling tower cells CT 2-1 and CT 2-2 are seismically designed to ensure that they are not adversely affected by a seismic event or by failure of other cooling tower cells. This design includes "breakaway" connections to the remaining cooling tower cells. These breakaway connections ..... are constructed by cutting the major wooden structural members connecting CT 2-2 to CT 2-3 and splicing them together with weaker materials that will separate in the event of significant seismic loading.

For cooling tower cell CT 2-1, the portion of the circulating water system piping that is in scope for 54.4(a)(2) is the carbon steel piping outside the tower that supplies water to the tower. This portion of the piping has the potential for spatial interaction with safety-related electrical equipment due to spray or leakage. This carbon steel piping is subject to aging management review as shown in Tables 2.3.3.13-B and 3.3.2.13-9. This carbon steel circulating water system piping transitions to fiberglass upon entering CT 2-1. The fiberglass circulating water piping has no license renewal intended function as discussed below. Therefore, fiberglass circulating water piping is not included in the LRA Section 3.3 tables.

The fiberglass circulating water piping is nonsafety-related and supports no system intended functions for 54.4(a)(1) or (a)(3). Pipe supports on this piping are part of the wooden tower structure and are subject to aging management review and included in the Structures Monitoring Program to ensure the piping cannot physically impact safety-related equipment. Following onset of the recent partial failure of CT 2-4, two lengths of the circulating water piping separated at a connecting joint. Failure of vertical wooden structural columns caused the piping to sag and separate at the joint. Managing the effects of aging on the wooden tower structure will prevent a similar piping separation at the joints in CT 2-1. The seismic analysis shows that the pipe stays intact during a seismic event. No other credible failure mechanisms can cause wholesale failure of the fiberglass piping. Postulated failures involving minor leakage from piping joints could spray or leak water on internal Cell CT 2-1 components. These components are designed for a wetted environment during normal cooling tower operation and as such would not be adversely impacted. As a result, the fiberglass piping cannot prevent satisfactory accomplishment of any of the functions identified in 10 CFR 54.4(a)(1) due to spatial interaction from spray or leakage, and is not in scope and subject to aging management review under 54.4(a)(2).

If the fiberglass piping were subject to aging management review, the aging management review results would be that there are no aging effects requiring management due to the high corrosion resistance of fiberglass which is composed of glass fibers. This is consistent with NUREG-1801, Volume 2, Line V.F-8 that lists no aging effects for glass piping elements in raw water.

The cooling tower basin has a storage capacity of 1.45 million gallons that is sufficient for seven days of ACS operation. The available capacity assumes that cooling tower cells CT 2-3 through CT 2-9 collapse during a seismic event resulting in an estimated 170,427 gallons of water (equivalent to the volume of all material in these cells) being displaced (lost). The evaluation does not credit the volume of water in basin below cooling tower cells CT 2-10 and CT 2-11. The basin below these two cells is shallow and the small volume of water is conservatively not credited for available capacity. Because the volume of the basin beneath cells CT2-10 and CT2-11 is not credited, a postulated collapse of the wooden structure of these two cells displaces no credited volume.

The potential for debris blockage of the ACS suction following an event involving collapse of cooling tower cells CT 2-3 through CT 2-11 has also been evaluated. The velocity through the suction grating at an ACS flow rate of 8000 gpm is 0.25 ft/sec which is 10% of the velocity required to keep sediment in suspension. This low velocity coupled with the tower cross bracing in two directions will prevent migration of debris to the ACS suction. The RHRSW system takes suction from a pit in the northwest corner of CT 2-1. The pit is approximately 60 feet from the nearest non-Seismic Class I cell. The suction pit is covered by steel grating. During alternate cooling system operation, RHRSW system flow is recirculated through CT 2-1. The only flow into CT 2-1 from the basin below the remaining cells is the flow required to make up for normal operating losses, such as, evaporation and drift. The flow rate from adjacent cells into CT 2-1 is low with a resulting velocity of less than a tenth of the 0.25 ft/sec velocity .... for flow through the grating over the suction pit.

Failure of cooling tower cells CT 2-3 through CT 2-11 (9 of 11 cells) and associated components has no impact on safety-related cooling tower cell CT 2-1.

Subpart II:

As identified in the VYNPS LRA, the aging effects on the cooling tower wooden structures are:

- (a)   change in material properties,
- (b)   cracking, and
- (c)   loss of material.

The aging mechanisms associated with the partial failure of CT 2-4 are:

- (a)   iron salt attack (formation of iron salts in the wood where ferrous hardware contacts the lumber and degrades the wood cells),
- (b)   fungal attack (wood destroying microscopic organism called decay fungi that forms in wood exposed to suitable temperature 40°F-140°F in moist environment), and
- (c)   repeated wetting and drying cycles causing wood checking and physical damage which reduces wood strength.

The circulating water piping within the cooling tower is made of fiberglass and is secured in wooden support saddles. The piping separation event resulted from the distribution deck sag that caused the bell/spigot joint to separate. It did not result from the effects of aging on the fiberglass piping. The wooden saddles supporting the distribution header were found in good condition with no significant degradation.

The supporting columns for the circulating water header experienced a reduction in strength due to iron salt attack and fungal attack at the upper spliced joints that caused cracking. This caused the initial failure of several support columns that led to deck sag and separation of the fiberglass circulating water piping joint, thereby increasing the local water loading, causing the additional column failures that lead to the partial failure of CT 2-4.

2-155

Response to Part C:

The circulating water piping separated due to the initial CT 2-4 column failure, rather than due to the effects of aging. This failure does not indicate a need to change the aging management programs for the circulating water piping. Thus, there is no impact on the aging management programs for circulating water piping.

Aging effects identified in the VYNPS LRA for the cooling tower structural elements are; loss of material, cracking and change in material properties. These aging effects are consistent with those associated with the failure of CT 2-4. The LRA identifies a need for enhancing the Structures Monitoring Program to add guidance for performing examinations of the wood cooling tower elements as appropriate to identify a loss of material, cracking, or change in material properties. This enhancement will include details for the examination and acceptance criteria for wood structures and structural components (i.e., columns and circulating water pipe supports) to ensure aging effects are identified and corrected prior to a loss of intended function. To detect a change in material properties, the enhancement will entail inspections that are more involved than remote visual surface inspections. Lessons learned from review of the failure of CT 2-4 will be considered in implementation of the enhancement identified for the Structures Monitoring Program.

The staff determined that the applicant has appropriately included cooling tower cells CT 2-1 and CT 2-2 within the scope of license renewal in accordance with the requirements of 10 CFR 54.4(a)(1) and (a)(2), respectively, and has committed (Commitment No. 21) to enhance and apply the Structures Monitoring Program to the cooling towers. In addition, the applicant has articulated the significant differences in design, material specifications, level of quality assurance oversight and maintenance between cooling tower cells CT 2-1 through CT 2-2 and those of cooling tower cell CT 2-3 through CT 2-11. These features, along with the execution of the Structures Monitoring Program, would preclude cooling tower cells CT 2-1 and CT 2-2 from failing in the manner of cooling tower cell CT 2-4. The additional information provided by the applicant demonstrated that cooling tower cells CT 2-3 through CT 2-11 do not meet the criteria of 10 CFR 54.4(a) for inclusion within the scope of license renewal in that they do not perform and intended function as defined by 10 CFR 54.4(a)(1) or (a)(3) and also that their failure would not prevent a safety-related SSC from performing its intended function as defined by 10 CFR 54.4(a)(2). Based on a review of the additional information provided by the applicant, the staff finds the applicant's response to RAI 2.4.4-2 acceptable.

2.4.4.3 Conclusion

The staff reviewed the LRA and related structural components to determine whether the applicant failed to identify any SSCs within the scope of license renewal or subject to an AMR. The staff finds no such omissions. On the basis of its review, the staff concludes that there is reasonable assurance that the applicant has adequately identified the process facilities components within the scope of license renewal, as required by 10 CFR 54.4(a), and those subject to an AMR, as required by 10 CFR 54.21(a)(1).

2.4.5 Yard Structures

2.4.5.1 Summary of Technical Information in the Application

LRA Section 2.4.5 describes the yard structures, structures not contained within the primary containment, reactor building, intake structure, or process facilities. Yard structures include the condensate storage tank foundation and enclosure structure, FO storage tank foundation and transfer pump house, N_2 storage tank foundation and enclosure, low-pressure CO_2 tank foundation and enclosure, JDD building, startup transformer foundation, switchyard relay house, trenches, manholes, duct banks, Vernon tie transformer foundation, Vernon Dam and hydroelectric station, and transmission towers. The condensate storage tank is near the southeast corner of the turbine building. The carbon steel enclosure houses safety-related equipment of the CST system. The FO storage tank holds make-up fuel for the EDG day tanks. A FO transfer pump house contains the FO pumps. The liquid N_2 storage tank enclosure is a seismic Class I structure designed so no instantaneous introduction of a high concentration of N_2 gas into the DG air intake occurs if the storage tank fails. A restraining wall around the base of the tank collects liquid N_2 and minimizes surface area to limit the boil-off rate of spilled N_2. The tank, located adjacent to the east side of the reactor building, is supported by a reinforced concrete foundation and structural steel support columns to meet seismic design requirements. The reinforced concrete CO_2 tank (TK-115-1) foundation is adjacent to the northeast corner of the switchgear room. A metal enclosure houses and protects electrical and mechanical equipment for the tank against the environment.

The JDD powers emergency lighting credited for alternate shutdown in the safe shutdown capability analysis. The start-up transformers (T-3A & B) on the west side of the turbine building are supported by reinforced concrete pedestals raised above a crushed rock bed. The startup transformers provide power during recovery from SBO. The switchyard control house, also known as the switchyard relay house, a single-story structure in the main switchyard, houses relays that control the offsite 115 kV lines. The trenches, manholes and duct banks throughout the VYNPS site, support and protect plant equipment. Those that support or protect equipment within the scope of license renewal are also in-scope. Duct banks route electrical cables between buildings and in the switchyard area.

The Vernon tie transformer is on a reinforced concrete slab located approximately 50 feet northwest of the west cooling tower and formed on a gravel and sand base to minimize frost heaving. The Vernon tie transformer is credited for SBO. Vernon Dam on the Connecticut River is constructed of concrete and steel and used for hydro-electric generation as an alternate source of AC power in an SBO. The dam and powerhouse are founded on compact rock and the power block superstructure is comprised of reinforced concrete, masonry brick, and structural steel. The dam is not a site structure owned by the applicant. Transmission towers are constructed of galvanized steel reinforced concrete foundations. In-scope towers are the 115 kV tower in the 115 kV switchyard, the 115KV angle tower located west of the turbine building, and the 115/345 kV shared tower in the 345 kV switchyard.

The yard structures have safety-related components relied upon to remain functional during and following DBEs. The failure of nonsafety-related yard structure SSCs potentially could prevent the satisfactory accomplishment of a safety-related function. In addition, the yard structures perform functions that support fire protection and SBO.

LRA Table 2.4-5 identifies the following yard structures component types within the scope of license renewal and subject to an AMR:

- steel and other metals
- concrete

The yard structures component intended functions within the scope of license renewal include the following:

- shelter or protection to safety-related equipment, including radiation shielding and pipe whip restraint

- protective barrier for flood events

- missile barrier

- structural or functional support to nonsafety-related equipment the failure of which could impact safety-related equipment

- structural or functional support for equipment required to meet fire protection, environmental qualification, PTS, ATWS, or SBO regulations

- structural or functional support for safety-related equipment

2.4.5.2 Staff Evaluation

The staff reviewed LRA Section 2.4.5 using the evaluation methodology described in SER Section 2.4 and the guidance in SRP-LR Section 2.4.

The staff evaluated the structural component functions described in the LRA and UFSAR to verify that the applicant has not omitted from the scope of license renewal any components with intended functions as required by 10 CFR 54.4(a). The staff then reviewed those components that the applicant has identified as within the scope of license renewal to verify that the applicant has not omitted any passive and long-lived components subject to an AMR as required by 10 CFR 54.21(a)(1).

In RAI 2.4.5-1 dated August 3, 2006, the staff stated that Table 2.4.5 lists "Vernon Dam external walls above/below grade" as a component, and "Vernon Dam external walls, floor slabs and interior walls" as another component. The staff requested that the applicant provide clarification for the two different components.

In its response dated September 5, 2006, the applicant provided the following response:

> In Table 2.4.5, item "Vernon Dam external walls above/below grade" refers to the outside surface of the exterior walls and the second line item "Vernon Dam external walls, floor slabs and interior walls" refers to the interior surface of the exterior walls along with floors and interior walls. This distinction is consistent with the treatment of each of these as having separate environments as shown in Table 3.5.2-5.

Based on its review, the staff finds the applicant's response to RAI 2.4.5-1 acceptable because it distinguishes the exterior surface of the Vernon Dam wall from the interior surface of the wall, which are subjected to different environments. The staff's concern described in RAI 2.4.5-1 is resolved.

2.4.5.3 Conclusion

The staff reviewed the LRA and related structural components to determine whether the applicant failed to identify any SSCs within the scope of license renewal or subject to an AMR. The staff finds no such omissions. On the basis of its review, the staff concludes that there is reasonable assurance that the applicant has adequately identified the yard structures components within the scope of license renewal, as required by 10 CFR 54.4(a), and those subject to an AMR, as required by 10 CFR 54.21(a)(1).

2.4.6 Bulk Commodities

2.4.6.1 Summary of Technical Information in the Application

LRA Section 2.4.6 describes the bulk commodities, structural components or commodities that perform or support intended functions of in-scope SSCs. Bulk commodities unique to specific structures are included in the reviews for those structures (SER Sections 2.4.1 through 2.4.5). This section addresses bulk commodities common to in-scope SSCs (e.g., anchors, embedments, pipe and equipment supports, instrument panels and racks, cable trays, and conduits).

The bulk commodities have safety-related components relied upon to remain functional during and following DBEs. The failure of nonsafety-related bulk commodity SSCs potentially could prevent the satisfactory accomplishment of a safety-related function. In addition, the bulk commodities perform functions that support fire protection, ATWS, SBO, and environmental qualification.

LRA Table 2.4-6 identifies the following bulk commodity component types within the scope of license renewal and subject to an AMR:

- steel and other metals
- concrete
- elastomers and other materials
- threaded fasteners

The bulk commodity component intended functions within the scope of license renewal include the following.

- shelter or protection to safety-related equipment, including radiation shielding and pipe whip restraint

- rated fire barrier to confine or retard a fire from spreading

- protective barrier for flood events

- insulation

2-159

- missile barrier

- pressure boundary

- structural or functional support to nonsafety-related equipment the failure of which could impact safety-related equipment

- structural or functional support for equipment required to meet fire protection, Environmental qualification, PTS, ATWS, or SBO regulations

- structural or functional support for safety-related equipment

2.4.6.2 Staff Evaluation

The staff reviewed LRA Section 2.4.6 using the evaluation methodology described in SER Section 2.4 and the guidance in SRP-LR Section 2.4.

The staff evaluated the structural component functions described in the LRA and UFSAR to verify that the applicant has not omitted from the scope of license renewal any components with intended functions as required by 10 CFR 54.4(a). The staff then reviewed those components that the applicant has identified as within the scope of license renewal to verify that the applicant has not omitted any passive and long-lived components subject to an AMR as required by 10 CFR 54.21(a)(1).

In RAI 2.4.6-1 dated August 3, 2006, the staff stated that Table 2.4.6 lists "Flood curbs" as a component with intended functions for flood barrier and shelter or protection, and another component "Flood curbs" with an intended function for flood barrier. The staff requested that the applicant provide clarification for the two different components.

In its response dated September 5, 2006, the applicant provided the following response:

> For VYNPS, flood curbs constructed of either concrete or steel perform the same intended function, which is to provide shelter or protection by serving as flood barriers. In essence, flood barrier and shelter or protection are the same function and both entries for flood curbs fulfill the same function.

Based on its review, the staff finds the applicant's response to RAI 2.4.6-1 acceptable because the applicant explained that the two entries for flood curbs perform the same intended function. The staff's concern described in RAI 2.4.6-1 is resolved.

2.4.6.3 Conclusion

The staff reviewed the LRA and related structural components to determine whether the applicant failed to identify any SSCs within the scope of license renewal or subject to an AMR. The staff finds no such omissions. On the basis of its review, the staff concludes that there is reasonable assurance that the applicant has adequately identified the bulk commodities components within the scope of license renewal, as required by 10 CFR 54.4(a), and those subject to an AMR, as required by 10 CFR 54.21(a)(1).

2.5 Scoping and Screening Results: Electrical and Instrumentation and Control Systems

This section documents the staff's review of the applicant's scoping and screening results for electrical and instrumentation and control (EIC) systems.

In accordance with the requirements of 10 CFR 54.21(a)(1), the applicant must list passive, long-lived SCs within the scope of license renewal and subject to an AMR. To verify that the applicant properly implemented its methodology, the staff's review focused on the implementation results. This focus allowed the staff to confirm that there were no omissions of EIC system components that meet the scoping criteria and subject to an AMR.

The staff's evaluation of the information in the LRA was the same for all EIC systems. The objective was to determine whether the applicant has identified, as required by 10 CFR 54.4, components and supporting structures for EIC systems that appear to meet the license renewal scoping criteria. Similarly, the staff evaluated the applicant's screening results to verify that all passive, long-lived components were subject to an AMR as required by 10 CFR 54.21(a)(1).

In its scoping evaluation, the staff reviewed the applicable LRA sections and component drawings, focusing on components that have not been identified as within the scope of license renewal. The staff reviewed relevant licensing basis documents, including the UFSAR, for each EIC system to determine whether the applicant has omitted from the scope of license renewal components with intended functions as required by 10 CFR 54.4(a). The staff also reviewed the licensing basis documents to determine whether the LRA specified all intended functions as required by 10 CFR 54.4(a). The staff requested additional information to resolve any omissions or discrepancies identified.

Once the staff completed its review of the scoping results, the staff evaluated the applicant's screening results. For those SCs with intended functions, the staff sought to determine: (1) if the functions are performed with moving parts or a change in configuration or properties, or (2) if they are subject to replacement based on a qualified life or specified time period, as required by 10 CFR 54.21(a)(1). For those that did not meet either of these criteria, the staff sought to confirm that these SCs were subject to an AMR, as required by 10 CFR 54.21(a)(1). If discrepancies were identified, the staff requested additional information to resolve them.

2.5.1 Summary of Technical Information in the Application

LRA Section 2.5 describes the EIC systems. Plant EIC systems are included within the scope of license renewal as are EIC components in mechanical systems. The default inclusion of plant EIC systems within the scope of license renewal reflects the method for IPAs of electrical systems. This method differs from those used for IPAs of mechanical systems and structures.

VYNPS electrical commodity groups correspond to two of the commodity groups identified in NEI 95-10: (1) high-voltage insulators and (2) cables and connections, busses, and electrical portions of EIC penetration assemblies. The IPA eliminated commodity groups and specific plant systems from further review as the intended functions of commodity groups were examined. In addition to the plant electrical systems, certain switchyard components required to restore offsite power following a SBO were conservatively included within the scope of license renewal although they are not relied on in safety analyses or plant evaluations to perform functions for compliance with SBO regulations. The offsite power system provides the electrical interconnection between the generator and the offsite transmission network and between the offsite network and the auxiliary buses as well as other buildings and facilities.

The EIC systems perform functions that support SBO.

LRA Table 2.5-1 identifies the following EIC systems component types within the scope of license renewal and subject to an AMR:

- cable connections (metallic parts)

- electrical cables, connections, and fuse holders (insulation) not subject to 10 CFR 50.49 Environmental qualification requirements

- electrical cables not subject to 10 CFR 50.49 Environmental qualification requirements used in instrumentation circuits

- fuse holders (insulation material)

- high-voltage insulators

- inaccessible medium-voltage (4.16 kV to 22 kV) cables (e.g., installed underground in conduit or direct buried) not subject to 10 CFR 50.49 Environmental qualification requirements

- switchyard bus

- transmission conductors

The EIC systems component intended functions within the scope of license renewal include the following:

- provide electrical connections to specified sections of an electrical circuit to deliver voltage, current, or signals
- insulate and support electrical conductor

2.5.2 Staff Evaluation

The staff reviewed LRA Section 2.5 and UFSAR Sections 7 and 8 using the evaluation methodology described in SER Section 2.5. The staff conducted its review in accordance with the guidance described in SRP-LR Section 2.5, "Scoping and Screening Results: Electrical and Instrumentation and Controls Systems." The staff reviewed the scoping methodology of the applicant, and considered it to be acceptable in accordance with the "Plant Spaces" approach method in NUREG-1800, Revision 1, Table 2.5.1. This approach eliminates the need for unique identification of every component and its specific location. This assures components are not excluded from an AMR.

As documented in SER, Section 3.6.2.3.1, the staff determined that uninsulated ground conductors are not in the scope of licence renewal and do not require an AMR.

In RAI 2.5-1, the staff requested the applicant to provide brief descriptions of the systems, listed in LRA Table 2.2-1b, explaining how each system serves one or more functions listed in 10 CFR 54.4(a).

In its response dated September 5, 2006, the applicant stated that:

> As described in LRA Section 2.5, all plant electrical and Instrumentation and Control (EIC) systems are included in the scope of license renewal. EIC equipment in mechanical systems is included in the scope of license renewal, regardless of whether the mechanical system is included in-scope. Including components beyond those actually required is referred to as an encompassing review. This method eliminates the need for unique identification of each system and its specific function. This assures components are not improperly excluded from the scope of license renewal.

Based on its review, the staff finds the above response to the RAI 2.5-1 acceptable because when used with "Plant Spaces" approach, this method eliminates the need for unique identification of each system and its specific function. The staff's concern described in RAI 2.5-1 is resolved.

In RAI 2.5-2, the staff requested the applicant to provide details of Vermont Yankee Nuclear Power Station's alternate alternating current (AAC) source, and also describe the offsite power recovery paths from switchyard to the onsite distribution which are in the license renewal scope to satisfy the requirements of 10 CFR 50.63.

In its response dated September 5, 2006, the applicant stated that:

> The parts of the AAC that are subject to AMR are explained in the response to RAI 3.6.2.2-N-08. The offsite power recovery paths from switchyard to the onsite distribution system which are in the license renewal scope are the source fed through the start-up transformers and a delayed access circuit from the 345 kV switchyard through the main and auxiliary transformers via the isophase bus. Specifically, the start-up transformer path includes; the 115 kV switchyard circuit breaker feeding the start-up transformers, the start-up transformers, the circuit breaker-to-transformers and transformer-to-onsite electrical distribution interconnections, and the associated control circuits and structures. The delayed access circuit is made available by opening the generator no-load disconnect switch and establishing a feed from the 345kV switchyard through the main and auxiliary transformers via the isophase bus.

The staff reviewed the applicant response to RAI 3.6.2.2-N-08, provided in the letter dated July 14, 2006, in which it stated that the VHS is the AAC source credited for Vermont Yankee Nuclear Power Station (VYNPS) to demonstrate compliance with 10 CFR 50.63, loss of all alternating current power (the station blackout rule). As such, all VHS structures, systems, and components (SSCs) are in the scope of license renewal.

Based on its review of the response to RAI 3.6.2.2-N-08, and further clarifications provided by the applicant in its letter dated January 4, 2007, Attachment 4, the staff finds the applicant's response to RAI 2.5-2 acceptable because the applicant has included all necessary components of the AAC source in the scope of license renewal. The staff's concern described in RAI 2.5-2 is resolved.

The applicant initially excluded metal-enclosed bus connections, and bus enclosure assemblies and insulators from the AMR. However, in its response dated September 5, 2006 to the staff's RAI 2.5-3, the applicant clarified that the metal-enclosed isophase bus is now included in the AMR. This isophase bus is part of the delayed access circuit (to support SBO recovery actions) from the 345 kV switchyard through the main generator step-up transformer and unit auxiliary transformer. The applicant stated that the VYNPS Metal Enclosed Bus Inspection Program will manage the effects of aging of the isophase bus and will be consistent with the GALL Report aging management program XI.E4 (NUREG-1801, Volume 2, Rev 1).

Based on above response provided by the applicant in its letter dated September 5, 2006, the staff considers that the applicant has included necessary components of the metal-enclosed bus connections, bus enclosure assemblies and insulators subject to an AMR. The RAI 2.5-3 response is considered acceptable. The staff's concern described in RAI 2.5-3 is resolved.

In RAI 2.5-4, the staff asked the applicant to provide justification, in detail, why the cable connections (metallic portion) was not included in the scope of an AMR although the GALL Report aging management program XI.E6, "Electrical Cable Connections not Subject to 10 CFR 50.49 Environmental Qualification Requirements," recommended such an aging managing program.

In its letter dated September 5, 2006, the licensee provided the following justification:

> Metallic parts of electrical cable connections that are exposed to thermal cycling and ohmic heating are those carrying significant current in power supply circuits. VYNPS power cables are in a continuous run from the supply to the load. The connections to the supply and to the load are parts of active components that are not subject to aging management review in accordance with 10 CFR 54.21. As discussed in the statement of considerations for the license renewal rule, maintenance rule activities are credited with managing the effects of aging on active components.
> The fast action of circuit protective devices at high currents mitigates stresses associated with electrical faults and transients. In addition, mechanical stress associated with electrical faults is not a credible aging mechanism because of the low frequency of occurrence for electrical faults. Therefore, electrical transients are not aging mechanisms.
>
> Metallic parts of electrical cable connections exposed to vibration are those associated with active components that cause vibration. Active components are not subject to aging management review in accordance with 10 CFR 54.21. As discussed in the statement of considerations for the license renewal rule, maintenance rule activities are credited with managing the effects of aging on active components.

Corrosive chemicals are not stored in most areas of the plant. Routine releases of corrosive chemicals to areas inside plant buildings do not occur during plant operation and corrosive chemicals are not a normal environment for electrical connections. Contamination of electrical connections causes rapid degradation independent of the age of the connection components. Corrosion due to contamination is due to the contamination event rather than aging. Therefore, chemical contamination is not an aging mechanism for electrical connections.

Corrosion and oxidation occur in the presence of moisture or contamination such as industrial pollutants and salt deposits. Enclosures and splice materials protect metal connections from moisture and contamination. Therefore, oxidation and corrosion are not applicable aging mechanisms.

Electrical cable connections at VYNPS are inspected in accordance with the maintenance rule program as directed by plant procedures. The maintenance rule program, based on industry guidance provided in NUMARC 93-01 and Reg. Guide 1.160, complies with 10 CFR 50.65. The maintenance rule program includes performance monitoring and trending. Monitoring and trending includes normal plant maintenance activities. Maintenance includes activities associated with identifying and correcting actual or potential degraded conditions (e.g., repair, surveillance, diagnostic examinations, and preventive measures).

Thermography is used to detect potential degraded conditions. Thermography can detect "hot spots" in cable connections that are indicative of a high resistance connection.

As a part of the maintenance rule program, periodic assessments are performed. A periodic assessment is performed to evaluate the effectiveness of maintenance activities. This assessment is performed at least every operating cycle, not to exceed 24 months. Plant operating experience has shown that the maintenance rule program has been effective at detecting, evaluating and repairing electrical cable connection degradation.

The maintenance rule program includes scoping, performance monitoring, trending and periodic assessments. This program provides reasonable assurance that electrical cable connections will remain capable of performing their intended functions through the period of extended operation. No aging management program (AMP) for license renewal is required at VYNPS since the regulatory mandated maintenance rule program effectively maintains electrical cable connections.

Subsequent to above response, on November 30, 2006, NEI held a meeting with NRC. Based on this meeting, XI.E6 program was revised to be a one-time inspection of a representative sample of cable connections subject to aging management review. In its letter dated January 4, 2007, Attachment 7, the applicant agreed to a plant-specific Bolted Cable Connection Program.

Based on licensee agreement to implement a Bolted Cable Connection Program (Commitment No. 42) as detailed in its letter dated January 4, 2007, the staff considers the issue raised in RAI 2.5-4 resolved.

2.5.3 Conclusion

The staff reviewed the LRA Section 2.5, the UFSAR, and the supplemental information provided by the applicant in its letters dated September 5, 2006, and January 4, 2007, to determine whether any SSCs that should be within the scope of license renewal or subject to an AMR had not been identified by the applicant. No omissions were identified. On the basis of its review, the staff concludes that there is reasonable assurance that the applicant had adequately identified the electrical commodity group components that are within the scope of license renewal, as required by 10 CFR 54.4(a), and that are subject to an AMR, as required by 10 CFR 54.21(a)(1).

2.6 Conclusion for Scoping and Screening

The staff reviewed the information in LRA Section 2, "Scoping and Screening Methodology for Identifying Structures and Components Subject to Aging Management Review and Implementation Results," and determines that the applicant's scoping and screening methodology was consistent with the requirements of 10 CFR 54.21(a)(1) and the staff's positions on the treatment of safety-related and nonsafety-related SSCs within the scope of license renewal and on SCs subject to an AMR is consistent with the requirements of 10 CFR 54.4 and 10 CFR 54.21(a)(1).

On the basis of its review, the staff concludes, pending resolution of Confirmatory Items 2.3.3.2a-1, 2.3.3.2a-2, 2.3.3.12-1, 2.3.3.13a-1, 2.3.3.13e-1, and 2.3.3.13m-1, that the applicant has adequately identified those systems and components within the scope of license renewal, as required by 10 CFR 54.4(a), and those subject to an AMR, as required by 10 CFR 54.21(a)(1).

The staff concludes that there is reasonable assurance that the applicant will continue to conduct the activities authorized by the renewed license in accordance with the CLB and any changes to the CLB in order to comply with 10 CFR 54.21(a)(1), in accordance with the Atomic Energy Act of 1954, as amended, and NRC regulations.

NRC FORM 335 (9-2004) NRCMD 3.7	U.S. NUCLEAR REGULATORY COMMISSION	1. REPORT NUMBER (Assigned by NRC, Add Vol., Supp., Rev., and Addendum Numbers, If any.)
BIBLIOGRAPHIC DATA SHEET *(See instructions on the reverse)*		NUREG-1907, Vol. 1

2. TITLE AND SUBTITLE

Safety Evaluation Report Related to the License Renewal of the Vermont Yankee Nuclear Power Station

3. DATE REPORT PUBLISHED	
MONTH	YEAR
May	2008

4. FIN OR GRANT NUMBER

5. AUTHOR(S)

Jonathan G. Rowley

6. TYPE OF REPORT

final

7. PERIOD COVERED *(Inclusive Dates)*

01/25/2006 - 03/20/2008

8. PERFORMING ORGANIZATION - NAME AND ADDRESS *(If NRC, provide Division, Office or Region, U.S. Nuclear Regulatory Commission, and mailing address; if contractor, provide name and mailing address.)*

Division of License Renewal
Office of Nuclear Reactor Regulation
U.S. Nuclear Regulatory Commission
Washington, D.C. 20555-001

9. SPONSORING ORGANIZATION - NAME AND ADDRESS *(If NRC, type "Same as above"; if contractor, provide NRC Division, Office or Region, U.S. Nuclear Regulatory Commission, and mailing address.)*

Same as item number 8 above

10. SUPPLEMENTARY NOTES
Docket No. 50-271

11. ABSTRACT *(200 words or less)*

This safety evaluation report (SER) documents the technical review of the Vermont Yankee Nuclear Power Station (VYNPS) license renewal application (LRA) by the United States (US) Nuclear Regulatory Commission (NRC) staff (the staff). By letter dated January 25, 2006, Entergy Nuclear Operations, Inc. (ENO or the applicant) submitted the LRA in accordance with Title10, Part 54, of the Code of Federal Regulations, "Requirements for Renewal of Operating Licenses for Nuclear Power Plants." ENO requests renewal of the VYNPS operating license (Facility Operating License NumberDPR-28) for a period of 20years beyond the current expiration at midnight March 21, 2012.

VYNPS is located approximately five miles south of Brattleboro, Vermont. The NRC issued the VYNPS construction permit on December 11, 1967, and the operating license on February 28, 1973. VYNPS is of a Mark 1 Boiling Water Reactor (BWR) design. General Electric supplied the nuclear steam supply system and Ebasco originally designed and constructed the plant. The VYNPS licensed power output is 1912 megawatt thermal with a gross electrical output of approximately 650 megawatt electric.

This SER presents the status of the staff's review of information submitted through February 21,2008, the cutoff date for consideration in the SER. The staff identified six confirmatory items which were resolved before the staff made a final determination on the LRA. SER Section 1.6 summarizes these items and their resolution. Section 6.0 provides the staff's final conclusion on the review of the VYNPS LRA.

12. KEY WORDS/DESCRIPTORS *(List words or phrases that will assist researchers in locating the report.)*

10 CFR Part 54, license renewal application, Vermont Yankee Nuclear Power Station, scoping, screening, aging management, time-limited aging analysis, TLAA, environmentally assisted fatigue, metal fatigue, safety evaluation report, Docket No. 50-271, operating license number DPR-28, Entergy Nuclear Operations, Inc., SER, VYNPS, flow-acclerated corrosion, FAC, CHECWORKS

13. AVAILABILITY STATEMENT

unlimited

14. SECURITY CLASSIFICATION

(This Page)

unclassified

(This Report)

unclassified

15. NUMBER OF PAGES

16. PRICE